A MOLECULE AWAY FROM MADNESS

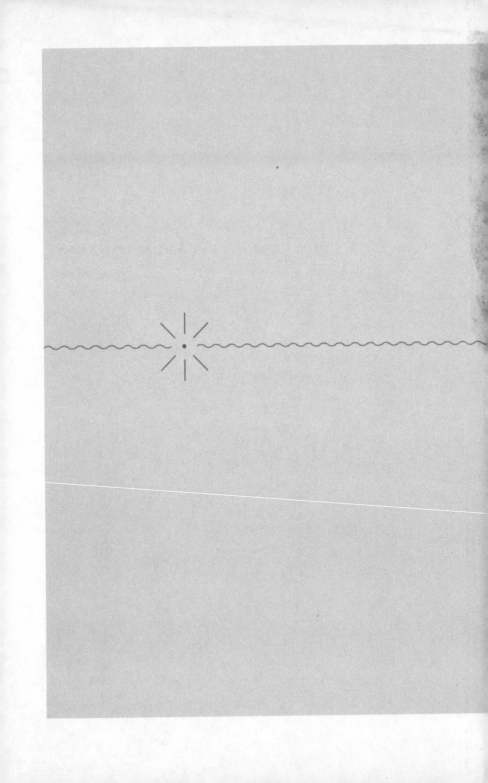

A MOLECULE AWAY FROM MADNESS

Tales of the Hijacked Brain

Sara Manning Peskin

W. W. NORTON & COMPANY
Independent Publishers Since 1923

A Molecule Away from Madness is a work of nonfiction. The names of patients and family members have been changed.

For information about permission to reproduce selections from this book, write to Permissions, W. W. Norton & Company, Inc., 500 Fifth Avenue, New York, NY 10110

For information about special discounts for bulk purchases, please contact W. W. Norton Special Sales at specialsales@wwnorton.com or 800-233-4830

Manufacturing by Lakeside Book Company
Book design by Lovedog Studio
Production manager: Lauren Abbate

Library of Congress Cataloging-in-Publication Data

Names: Peskin, Sara Manning, author.
Title: A molecule away from madness : tales of the hijacked brain / Sara Manning Peskin.
Description: First edition. | New York, NY : W. W. Norton & Company, [2022] | Includes bibliographical references and index.
Identifiers: LCCN 2021041595 | ISBN 9781324002376 (hardcover) | ISBN 9781324002383 (epub)
Subjects: LCSH: Neurosciences. | Brain—Research.
Classification: LCC RC327 .P47 2022 | DDC 612.8/2—dc23/eng/20211004
LC record available at https://lccn.loc.gov/2021041595

W. W. Norton & Company, Inc., 500 Fifth Avenue, New York, N.Y. 10110
www.wwnorton.com

W. W. Norton & Company Ltd., 15 Carlisle Street, London W1D 3BS

1 2 3 4 5 6 7 8 9 0

For Jeremy, who taught me to tell a story, and for JJ and Oliver, our captive audience

Contents

A MOLECULE AWAY FROM MADNESS

INTRODUCTION

AT YOUR BEGINNING, A TADPOLE-SHAPED CELL found the opaque edges of a human egg and burrowed inside. The fertilized egg—now an embryo—cinched itself and divided into two. Two cells became four, four became eight, and so on until something astounding happened: instead of remaining identical, each cell took on a different role.

Some cells were shipped off to the frontier to become skin. Others began manufacturing hormones that could make you sleepy, hungry, or nervous. Still others became muscle cells that could manipulate the bones in your growing skeleton.

The organ that defines your personality—the one that makes you, you—started off in the embryo as a sheet of cells roughly the size of a pencil tip. Over just a few days in early development, the sheet rolled up into the shape of a long tube. One end of the structure stretched to form your spinal cord, while the other blossomed into the very brain you are using to read this page today.

Just above your eyes, you developed neurons that help you control impulses. Neurons on the sides of your brain learned to interpret language and music. Toward the top of your head, neurons became specialists in arithmetic and judgment. Underneath, a set of neurons sorted out visual information couriered from the back of your eyeballs.

Voilà. You became the owner of the most complex machine known to humankind. Your brain has more than eighty-six billion neurons—more than the brain of any other animal on earth. It is larger in size than the brain of any other primate, and it holds more data than the most cutting-edge smartphone. Parts of our brains are so complex that they do not even develop fully until we reach our mid-twenties.

And yet.

Our brains have an Achilles' heel. The very molecules that make our brains work can also co-opt our personalities and destroy our ability to think. Our temperament, memories, and relationship to reality can all be lost to molecules that are billions of times smaller than our brains. Tales of guerrilla warfare have fascinated humans for millennia, but few of us realize that our brains are engaged every day in the same genre of conflict. We are forever surviving on the brink, battling molecules that can destroy our minds.

Molecule is a daunting word that turns out to have a simple meaning: a molecule is a group of atoms bound together. You are probably already familiar with atoms, like oxygen, carbon, and hydrogen. When atoms are linked together, we call the resulting structure a molecule.

Water is a molecule that contains two hydrogen atoms and one oxygen atom and is thus called H_2O. Thiamine—another molecule that will become important in this book—is also made of hydrogen and oxygen atoms, but it contains carbon and nitrogen atoms, too. Deoxyribonucleic acid (DNA) is an enormous, stringy molecule built from the same atoms as thiamine, with the addition of phosphorus.

All these molecules are so small that you cannot see them with traditional microscopes. A cup of water contains a septillion water molecules—more than a trillion times the world population. A grain of sand contains more molecules than there are insects on the earth. Even DNA, the largest molecule in the human body, is so small that scientists can directly visualize its structure only with a specialized microscope—and they just discovered how to do this in 2012.

But the size of molecules says nothing about their ability to co-opt the mind. This book is about molecular villains, millions of times smaller than the brain yet so nimble in hijacking its function. Scientists have written volumes about each of these molecules, but I like to think of them more casually as *mutants*, *rebels*, *invaders*, and *evaders*.

Mutants are altered sequences of DNA. If you consider DNA as a giant three-dimensional computer code, mutants are like small typos that cause the program to self-destruct. As you will see in the first chapters of this book, mutants can inflict deadly cognitive maladies on generation after generation—a sentence we are now close to commuting thanks to some of the most striking discoveries in all of neurology.

Rebels are aberrant proteins. Under normal circumstances, proteins are wildly talented molecules that carry out the directions given by our DNA. If we return to the idea of DNA as a computer code, proteins are the people and infrastructure that bring the code to life, like the conductors who operate trains according to the timetable dictated by an algorithm. But proteins can also rebel against us, taking aim at our brains and inflicting a rapid and dramatic destruction. Recalcitrant proteins can cause us to hallucinate, to erupt in anger, and to descend into a trembling dementia—all phenomena you will get to know well in the second part of this book.

Finally, there are so-called small molecules, much tinier than DNA and proteins, that can invade our brains when they are unwelcome, or be absent when we need them to be present. Returning to the locomotive analogy, you can think of small molecules either as obstacles that block the train tracks (invaders) or as fuel required for the train to move in the first place (evaders). As you will discover in the final chapters of this book, these tiny invaders and evaders can throw us into rapturous rages, turn us into constant liars, and instill an insidious, striking state of confusion.

The characters and conundrums that occupy the upcoming pages are not simply the stuff of scientific oddity. The stories in this book represent the foundation for the most exciting frontier in cognitive neurology today. By examining the molecules that hijack the brain, we can start to understand how we will treat Alzheimer's disease and other common brain diseases in the future.

Cancer treatment has undergone a revolution in the past twenty-five years, because researchers have identified the molecular causes of oncological diseases and designed molecular solutions. Likewise, molecular neurology is the answer to the common cognitive ailments that continue to plague our brains. The researchers who solved the mysteries that unfold in the next pages have laid the groundwork for neurology to go the way of oncology. Sometimes outlandish, often criticized, and forever devoted to their art, these scientists and doctors have brought cognitive neurology to where we stand today: on the precipice of a molecular breakthrough.

～～～～

MY OWN LOVE AFFAIR with molecules began in college, when I fumbled with pipettes and test tubes to learn how bacteria stitch together armor that protects them from antibiotics. I worked in a bustling laboratory with rows of black-topped research tables. Wooden desks, each assigned to a student, were blanketed with scientific papers, textbooks, and disintegrating coffee cups. Family photos tacked onto corkboards reminded people of the world outside.

Our team survived on a feeling of wonder for the minuscule. At one end of the room, a witty woman from Queens discovered how specialized molecules help bacteria to divide in half without exploding. In another corner, a shy and persistent woman recreated an ornate molecular complex in a test tube. A few desks over, a young father from Singapore figured out how bacteria build a molecule that makes them more resistant to antibiotics.

I went on to attend medical school and train in neurology. I became a dementia doctor, at once horrified and fascinated by the way Alzheimer's disease and other types of dementia can change a person's personality. Today, I spend most days watching my patients slowly disappear as their husbands, wives, children, and sometimes parents look on in anguish. I talk to patients who see people and animals that do not exist. They wake up in the middle of the night and ask their partners, "Why is that man sitting at the end of our bed?" or "Why is that rabbit staring at you?" I interview spouses who had been doting and devoted for decades but, as part of their dementia, turned to extramarital affairs and public displays of nudity. I am, in many ways, a guide on the path to nothingness.

As with a slowly sinking vessel, there are moments when patients and personalities bob to the surface, revealing a glimpse into a longed-for life. A caregiver will describe a patient's transient joy in hearing about the birth of a grandchild, before forgetting that the baby is related at all. A spouse will tell of a partner's sudden ability to offer comfort and empathy— an unexpected reversal of roles where the caregiver briefly becomes the cared-for. But eventually, many of my patients are simply gone, lost to Alzheimer's disease and other cognitive scourges, all caused by lethal molecules that we do not yet know how to defeat.

I am immersed daily in the real-life details of the crumbling mind, but the importance of single molecules is just as profound in my work now as it was when I spent my days in a basic science laboratory. Most of my patients are incurable precisely because we do not have a molecular solution to their ailment. When it comes to the most common cognitive calamities, we have yet to accomplish what cancer doctors achieved for their patients a quarter century ago.

This book tells the stories of patients whose lives have been upended by mutants, rebels, invaders, and evaders. It depicts the triumphs and failures of the scientists and doctors who devoted their careers to uncovering the secrets of the molecules that hijack the brain. These are tales of havoc—wild personality changes, memory loss, death, and afflictions in between—that illustrate what any neurologist knows, and what the people in these accounts have come to understand intimately: we are each just a molecule away from madness.

Part One

DNA
MUTANTS

DNA HAD A LACKLUSTER START IN THE SCIENTIFIC world.

The story begins with Dr. Friedrich Miescher, a near-deaf Swiss doctor who relegated himself to the laboratory in the mid-1800s after he realized he could no longer hear his patients. Miescher became an enraptured researcher, known for using his home chinaware set when he ran out of laboratory equipment, and for leaving his fiancée waiting at the altar while he finished an experiment (she went on to marry him anyway). Enthralled by the chemistry of pus, Miescher would collect used bandages from a nearby hospital and scrape their white contents into beakers that he stored across his laboratory. Contemporary accounts suggest he was unfazed by the origins of his substrate; he complained only that he was not able to acquire larger, fresher quantities of pus despite his best efforts.

In examining the pungent samples, Miescher found something he did not expect: in addition to molecules that earlier scientists had written about, cells in the pus also contained a stringy material that was rich in phosphorus atoms. Miescher had never read of anything like it before. He was not sure what it was doing in his cells. As far as he could tell—and he would turn out to be right—he had discovered something novel.

Miescher published a description of the curious substance in a scientific journal later that year. The paper was dry and verbose, taking up twenty pages that quickly provoked more scorn than praise. Some scientists thought the mystery molecule was simply a contaminant that Miescher had accidentally introduced into his experiments. Others, suspecting something more nefarious was at play, called his scientific integrity into question. Even people who thought his protocols were sound did not believe he had discovered the molecule that transmitted characteristics from generation to generation. At the time, Miescher himself thought the molecule was too chemically simple to contain instructions for building and running the diversity of living things on the planet.

Miescher's threadlike isolates soon became known as deoxyribonucleic acid, or DNA for short, but few people conceived that it had any relevance to heredity.* So for the next eighty years, DNA was nearly forgotten. Scientists focused on proteins, the diverse and surprisingly efficient molecules that carry out the grunt work of maintaining cellular life. It was only logical, researchers believed at the time, that such an astoundingly capable molecule as protein would also be

---

* The term *deoxyribose* refers to the chemical structure of DNA, which contains a sugar molecule, named ribose, that has lost one of its oxygen atoms. The term *nucleic* refers to a subcompartment of cells called the nucleus, where DNA is located. The addition of *acid* clarifies that DNA is slightly acidic, meaning that it tends to give off hydrogen as it is synthesized.

the substance that allowed traits to weave their way through lineages. Proteins were paramount, scientists thought; everything else was just schmutz.

The narrative changed only in 1944, thanks to Dr. Oswald Avery. Avery was a nearly retired Canadian bacteriologist with a narrow chin and a wide forehead that looked as if the top of his skull had stretched to accommodate the bulk of his brain. He was a drably dressed creature of habit, working out of an undecorated kitchen turned laboratory space at the Rockefeller Institute in New York City.

Like Miescher, Avery had trained as a physician and then left clinical medicine—in Avery's case, after feeling powerless to treat patients suffocating from lung disease. Turning to the research bench, he sought to understand more about the strange behavior of one of the most common pulmonary scourges, a bacterium called pneumococcus.

One of Avery's predecessors had discovered that pneumococci had a remarkable ability to learn new tricks. In the scientist's hands, innocuous strains of the bacteria could become infectious simply by mixing with the remains of destroyed, infectious bacteria. It was like learning to play guitar like Jimi Hendrix by hanging around the dead musician's grave. It was also, Avery realized, akin to parents passing on traits to their children.

Avery wanted to understand how bacteria could pick up new traits from their environment—how they could transform from being harmless to being infectious. To find the answer, he grew two flasks full of bacteria. In one of the containers, he cultivated an infectious population of pneumococci. In the other, he grew a noninfectious form of the same bacteria. Initially, he recapitulated his predecessor's work, killing the infectious bacteria and proving that something in its soupy debris could teach the noninfectious bacteria how to become virulent. Then, he began using the process of elimination to figure out what molecule enabled the phenomenon.

To test whether or not proteins were important to the experiment, Avery added a chemical that destroyed proteins in the infectious bacteria. To his surprise, the step had little effect on the outcome of his procedure. The harmless bacteria still learned to become infectious. Contrary to the prevailing scientific belief, proteins were not the critical molecule of heredity everyone had imagined.

Then Avery tried destroying the DNA in the debris from the infectious bacteria. Like an assembly line missing a component, the experiment stopped working. The innocuous bacteria could no longer learn to become dangerous. It had been DNA—and not protein—all along that allowed bacteria to pick up new skills from their environment. The experiment illustrated for the first time that DNA was the long-sought-after molecule that conferred hereditary traits. Nearly a century after it was discovered, scientists finally recognized DNA as the molecule that made children similar to their parents.

We now know that a complete copy of your DNA is present in almost every cell in your body—notable exceptions being red blood cells, which die without replicating, and sperm and egg cells, which have only half the usual complement of genetic information. But inside nearly every other cell, your DNA is divided into forty-six pieces called chromosomes, each built from millions of nucleotides.

If you consider the entire sequence of human DNA as a book, chromosomes would be chapters and nucleotides would be letters. Instead of having twenty-six letters, as in the English alphabet, human DNA is made of just four nucleotides: adenine, thymine, guanine, and cytosine—conveniently abbreviated A, T, G, and C. With only four nucleotides to use as building blocks, it is no wonder Miescher had doubted that DNA could be the molecule of heredity. How could a substance with so few parts encode enough information to account for the extraordinary variability of humans, plants, and animals living on the earth?

What Miescher did not know—what scientists would not learn for another century—was that the sequence of human DNA in each of our cells is nearly three billion nucleotides long. Stretched end to end instead of coiled tightly in your cells, the DNA in your body could reach to the sun and back many times over. Humans genetically differ from one another not because we have so many different nucleotides in our DNA, but rather because the nucleotides are strung together into a code that is so gargantuan that there are a seemingly infinite number of places where the sequence can differ from one person to another.

Most of the time, variations in DNA have no deleterious effect. You might have an A nucleotide at a certain place in your DNA code, whereas your neighbor might have a T at that location, and neither of you will experience any negative ramifications from the difference. In this way, our DNA is remarkably resilient. We can withstand a surprising number of mutations without experiencing any harm.

But sometimes, in particularly important parts of our DNA, even a change in a single nucleotide can be lethal. For families who unwittingly pass on these dangerous mutations, DNA can be the source of centuries-long torture, sewing the threads of catastrophe through vast networks of relatives. DNA, usually a spring of immense power, can instead become a fount of destruction.

Harmful DNA mutations can devastate any part of the body, but nowhere is the effect more striking than in the brain. In other organs, DNA mutations can cause us pain, disfigurement, and even death, but they do not upend the personalities that define us as individuals. In the brain, DNA mutations can rob us of empathy, memory, language, and other critical parts of our identities. The mutations can create a person wholly unlike the one our family and friends have come to know.

Our understanding of genetics has now become so extensive that we can sometimes identify people who will develop brain maladies even before they show symptoms. We can predict the future in a way that was never possible in the past. In some cases, the knowledge has even allowed scientists to intervene early enough that people never fall victim to the genetic curses woven into their DNA. Patients who were once untreatable are now curable.

So this is where we begin: with the molecule that defines us from birth, and with the scientists who are learning to protect our brains from our own DNA.

Chapter One

In Suspension

IN THE WAITING ROOM OF A HUNTINGTON'S DIS-
ease clinic, limbs writhed. Fingers curled. Legs that should
have rested on seat fabric instead lifted upward, twisting in the
air. Chairs jostled against the floor.

Amelia Ellman sat still, save for feet that vibrated with anxiety. She had not come to the clinic for a doctor to check her
symptoms; her muscles and mind still worked as well as any
other twenty-six-year-old's. Amelia had come to the office to
hear the results of her genetic testing, to have her fortune read
off a piece of paper sent from a laboratory that had analyzed
her DNA.

Amelia's mother had died of Huntington's disease the
previous year. Her demise had been slow and excruciating, unfolding over a decade. She had become irrational,
demented, and exhausted from unintentional movements
that made her limbs look like they were infused with a fluctuating electric current.

Over the same time, Amelia's movements had become only
more exact. Amelia had become an aerialist, dependent on

precise motions to keep herself afloat at gravity-defying angles. She could perform with balletic grace while hanging ten feet off the ground, suspended only by two reams of silk draped from a tall ceiling. She could swing in and out of a large hoop while it spun around dizzyingly in midair.

In the waiting room, where the ceiling was low and most people's gesticulations were jerky and unwanted, Amelia prepared to find out whether her career as a high-flying gymnast would be replaced with the shackles of wheelchairs and hospital beds. She had known for years that she had a 50 percent chance of inheriting the gene that causes Huntington's disease. With the results of her genetic testing, the statistics would instantaneously change. The odds of dying like her mother would be all or nothing. In the bland consultation room beyond the check-in desk, she would be released from uncertainty.

Amelia was not alone in her wait. Her grandmother, a nurse, sat in the chair next to her. Amelia's grandmother was resilient and nostalgic, a family historian intent on stuffing photo albums with glossy snapshots that often hid reality. It was she who had called from the nursing home when Amelia's mother died. She had delivered the news softly, knowing both women felt the same combination of sorrow and relief.

A year later, ready to receive Amelia's genetic testing results, the women made their way to a brick building on the corner of a small commercial street, across from an antique store and a hipster coffee shop. They rode the elevator to the fourth floor, then stepped into the waiting room of worst-case scenarios. Amelia gave her name at the front desk.

Soon, the doctor called for her. Without speaking a word, she and her grandmother stood up and walked past the receptionist, into the consultation room.

～～～～

AMELIA HAD LONG CONVINCED herself that the result from the laboratory would dictate bad news. *Just look at how horribly things have gone for me until now*, she thought. She saw her life as a series of blunders and near catastrophes. Her parents had divorced when she was three years old. In the wake of the separation, her mother had struggled to keep low-paying jobs. They had depended on sales at the local grocery store and often wore clothing collected from shelters.

In elementary school, Amelia had survived on visits from her grandparents. With their arrival, the street would be abuzz as her grandfather revved his motorcycle. Her grandmother would burst through the door armed with photographs of relatives whom Amelia never saw anymore. It was her grandparents who had initially patched the holes in her mother's finances, stepping in with their meager resources until they became overwhelmed with debt themselves.

By the time Amelia was twelve, rent checks had gone unpaid for months in a row. The landlord knocked on the door and, apologizing, asked them to leave. Amelia's next home was in a trailer park where she woke up at four o'clock on weekday mornings to take three buses to school. Exhausted, she rode the same route back each afternoon, arriving home just in time to go to sleep and do it all over again the next day. When the mobile home community was robbed one evening, Amelia and her mother moved a third time, now to a dingy apartment. Amelia began wondering how long they would have a place to stay at all. Then her curiosity was answered: they moved out of the apartment and into a motel.

Around the same time, Amelia noticed her mother's body was changing. Her arms wriggled and writhed with no pattern or predictability, as if controlled by a drunk puppeteer. Her hands would knock against chairs and tables, slowly shifting the furniture around the room. The sequence of sounds became familiar: Scratch. Profanity. Bump. Profanity. Screech. Profanity. Cooking turned into a cacophony of clanging silverware and pots. Sometimes the movements would become so profound that Amelia's mother would tumble to the floor, looking up at the ceiling as Amelia crouched over her face and tenderly grabbed her hands to help her back onto her feet.

Amelia was perplexed by the movements, but her mother had an inkling of what was happening; she had been adopted at birth from a biological mother about whom little information was known except that the woman suffered from Huntington's disease. Amelia's mother, like Amelia, had lived much of her early life under the threat of hereditary devastation.

As her mother's condition worsened, Amelia was largely left to raise herself. She found companionship in alcohol and prescription pills. She spent nights on friends' couches or on sidewalks, immersed in exhilarating highs and demonic lows. She dropped out of high school.

Then one day, at age sixteen, Amelia woke up in their motel room, makeup smudged across her face after a night that— like many others—she could not remember. She stared at her mother sitting at the kitchenette counter, smoking a cigarette, phone pressed to her ear as she begged a long-lost relative to send money.

Amelia considered the scene, a now-familiar picture. In a moment of self-preservation, she slipped out the door and onto the motel's balcony. She paced up and down the concrete walkway, clutching her cell phone as she found the number for child services and called to ask for help.

When social workers arrived at the door, Amelia's mother exploded with maternal ferocity. She pleaded to keep her daughter at home, offering to do whatever it took to appease the unwanted visitors. But the apartment was in disarray, and the report Amelia had provided over the phone—an exchange her mother would never learn about—was too alarming for the social workers to oblige. Stuffing her belongings into a garbage bag, Amelia looked back at her single-room home and took stock. Minutes later, she watched the motel sign shrink into the distance as she stared out the rear window of the social worker's car.

Amelia spent the next year living in a homeless shelter for children. She went to Alcoholics and Narcotics Anonymous meetings until she started to believe in her ability to withstand the draw of substances. Eventually, she found work in a Hello Kitty store at the local mall. She signed on for another job during her off-hours.

Clean and employed at age seventeen, Amelia took a bus across town to visit her mother for the first time in a year. She knocked on the peeling motel door, but nobody answered. She turned to leave, then noticed a thin woman sauntering down the path, tilting uneasily from one side to another. The figure's too-big shorts revealed wiry thighs. Bony hands grew off tiny wrists attached to stick-like arms that emerged from the woman's sleeves. Her feet, clad in sandals, slapped the ground when she walked.

Then there was the black purse with a long strap, an item Amelia knew well. It swung in grand sweeps as the woman made her way clumsily toward the motel door where Amelia was standing. There was recognition, then a hug that shocked Amelia as she felt just how much of her mother's body had disappeared while they had been apart.

In an unspoken instant, Amelia became her mother's caregiver. She used the money she had saved from her jobs to pay off the remaining two thousand dollars of the motel bill. She found a one-bedroom apartment in the building next door and lugged her mother's belongings there. She brought her own things, too—a mass of clothing and practical items she had acquired in the year since she had packed her belongings in a garbage bag under a social worker's gaze.

Amelia cut her mother's hair, which had become so matted that it crunched as the scissors made their way through it. She learned to bathe her mother: to perch her on the toilet seat, undress her, and move her carefully, one foot at a time, into the tub as her limbs banged against the plastic. She learned to follow her mother's increasingly rigid requests, like drying her ears out immediately after taking her out of the bathtub. "Ow, ow, ow," her mother would say, as if the gentlest handling was aggressive.

Amelia bought a mattress and put it on the floor for her mother to lie on during the day, since the writhing often launched her off of the couch without warning. Each morning, before leaving for work, Amelia would set out a can of Coca-Cola with a straw for her mother to drink. "Bring it closer," her mother would command, unsatisfied. Then, when Amelia obliged, her mother would soften. "I want you to take care of me forever," she would say. "I never want to go to a nursing home."

As one year and then most of another passed, Amelia worried about her mother spending hours alone while she was at work. *An accident is inevitable,* Amelia began to think. She called nursing homes looking for an opening, but every administrator who answered said she could not force her mother to move unless there was proof of immediate danger.

Soon, the risk became clearer. Increasingly shaky, her mother dropped a cigarette on herself and was lit on fire while Amelia was at work. By the time Amelia arrived at the hospital, parts of her mother's leathery skin, so intimately known through hours of bathing and dressing her, had turned black and blistery. As her mother cried out, "I want to go home," Amelia explained in guilt-laden words that she could not do what her mother asked. Instead, at age forty-one, her mother left the hospital for a nursing home.

Within the year, Amelia's mother was swallowed into an imaginary universe. She became convinced she owned a Walmart and that she and Amelia were still living together. "Here's my husband," she would say to the staff, pointing to nothing in particular. Between shifts at work, Amelia tried taking her mother outside. She spent hours interpreting her mother's grunts for the people who took care of her bodily needs, as if the two women were the last speakers of a dying language.

Eventually, even Amelia could no longer understand her mother's sounds. Without control over her movements, her mother was left with little ability to express anything—no chance to explain if she was hungry or tired, or if she wanted someone to change the channel on the television. Her throat could no longer open and close at normal intervals. Water, juice, and food made their way into her lungs. Coughing fits ensued, and then bouts of pneumonia.

In the summer of 2017, Amelia's mother died. Amelia cupped her hands over her eyes and cried, not for the death itself—which by then was expected—but for the tragedy of the downfall. She wailed for the finality of it all, for the realization that she would never truly know her mother.

Talking by phone, Amelia and her grandmother launched into the logistics of death. The money Amelia's mother had received from the government had more than covered the cost of her nursing home, leaving three hundred dollars to spare by the time she died. After hanging up, Amelia called a funeral home and arranged for her mother's cremation—a service that cost three hundred dollars. In death, her mother had unknowingly taken care of her own finances.

Entrenched in the same anxiety that had struck her mother—a feeling she now learned could be a predecessor to the movement symptoms of Huntington's disease—Amelia once again decided to change course. She wanted to know if she had inherited her mother's genetic mutation. If she had, she would prepare her mind and her muscles to be the strongest they could be when the disease eventually took over. She would travel, witnessing the world while she could still talk to strangers and board buses. Perhaps, she thought, she would adopt children.

Pacing around her apartment one morning, Amelia spoke to a nurse at a Huntington's disease clinic. She described images from her mother's demise and explained why she wanted to be tested for the gene that haunted her. She went into a clinic and talked with a psychiatrist, a neurologist, and a genetic counselor. She spat into a clear plastic tube and watched as the saliva was tucked away into an envelope to be sent to a laboratory in another state, where her future would be decoded.

Weeks later, Amelia and her grandmother crossed the threshold into a small consultation room at the clinic. Each woman sat down in one of the unmatching chairs. With little preamble, the verdict was delivered.

"You have it," the doctor said.

THE MOLECULAR TEST THAT Amelia took was developed using one of the most fortuitous discoveries in medical history. By the mid-twentieth century, the fledgling field of molecular genetics had stalled: researchers had known for two decades that DNA dictated hereditary traits, but they could not figure out which genes inflicted what diseases. The human genome was essentially a black box.

In the case of Huntington's disease, researchers could not find where the Huntington's disease gene—the piece of DNA that, when mutated, causes the disease—was hidden within the human genome. Scientists hoped that finding the gene would allow them to discover a treatment for the disease, but the search had proved to be excruciating. At the time, there were few tools to locate a particular sequence of DNA within the three-billion-nucleotide-long human genetic code. It was like trying to find a winning lottery ticket buried in two tons of garbage, with only your bare hands to help.

Then, in 1968, a twenty-three-year-old woman named Nancy Wexler learned that her mother had Huntington's disease. "It's a progressive, degenerative, neurological illness," her father explained as she sat on the living room couch at his apartment in Los Angeles. An enormous painting of Humpty Dumpty hung on the wall, the egg drawn with a perverse smile. "You have a fifty-fifty chance of inheriting the disease yourself," her father said.

Grappling with the news, Nancy Wexler made two decisions. The first, which she arrived at within the day, was that she would not have children. The second, which evolved slowly over the next months, would eventually make her famous: she decided to look for a cure for Huntington's disease.

Wexler hosted a series of workshops to engage scientists in the problem of Huntington's disease. The meetings were carefully orchestrated, geared toward early-career scientists in particular in order to limit egos and foster creativity. Wexler banned the use of prepared slides, hoping it would force participants to step out of their scientific niches and see the problem with fresh eyes.

In 1979, scientists at one of the workshops proposed a plan for finding the Huntington's disease gene: they would look at the DNA sequences nearby. Imagine a string of beads that you cut at a random point, yielding two separate fragments. Statistically, beads that start off close to each other on the strand are more likely to end up in the same piece after the string is severed; beads that begin near each other will probably stay that way.

The same is true of our DNA. Nucleotides that are close together on the same chromosome tend to be inherited together from one generation to the next. They are linked. Thanks to this phenomenon, scientists could use the DNA around the Huntington's disease gene as a proxy for the gene itself. Instead of trying to characterize a specific gene, they could settle for studying the right region.

On the surface, finding the DNA around the Huntington's disease gene would be just as difficult as finding the gene itself. Scientists did not have tools to look for either sequence directly. But at the workshop in 1979, the researchers discussed a solution. They would take advantage of tiny molecular scissors that could differentiate short sequences of DNA around the Huntington's disease gene. The scissors would cut up samples of a person's DNA in such a way that fragments close to the Huntington's disease gene would be a different length in people with Huntington's disease than in those without it.

The results would be family specific; in one clan, the procedure might yield a five-hundred-nucleotide-long piece of DNA in those with Huntington's disease and a two-hundred-nucleotide-long fragment in people not destined to get the disease. In another family, the same procedure might yield pieces that were three hundred nucleotides versus six hundred nucleotides. The important thing was that, within a single family, the length of the DNA fragment would correlate with whether or not the person had a normal Huntington's disease gene or one that would cause the affliction.

Then came the most unpredictable step. To visualize the relevant pieces of DNA—to see if the fragments differed in length between those with and without the disease—the researchers needed a tag that stuck within thirty million nucleotides of the gene, a distance covering just 1 percent of the human genome. At the time, fewer than twenty such tags existed worldwide. The odds of any of them working were dismal.

"It would take more than a decade" to create enough tags, one meeting attendee warned Wexler. "We'd be giving families false hope if we told them we were looking for the gene like this," another participant said. Pandemonium erupted at the workshop as scientists skeptical of the idea raced to the front of the room to illustrate why it was preposterously inefficient. The sound of chalk tapping feverishly against the blackboard was drowned out only by voices of exasperation.

Then a more optimistic view arose. "Whenever a new [tag] comes out, we'll just try it and see if it works," a scientist named David Housman said to the group. He thought the approach was more promising than others believed. He argued there was no reason to wait before trying the protocol with the few tags that already existed.

As the workshop came to a close, with some participants still outraged, Wexler offered Housman a small amount of money to begin the project. He in turn recruited Jim Gusella, a young Canadian geneticist who had no idea the venture would earn him the nickname Lucky Jim.

But it did. Just three years after the team began working on the project, Gusella sat in his office, mouth agape, marveling at the results. What cynics had said would take more than a decade had taken less than half that. By chance, one of the few tags already in existence happened to stick within five million nucleotides of the Huntington's disease gene. The scientists had found the chromosome, and even the section of the chromosome, where the Huntington's disease gene was located. It was a breakthrough for Huntington's disease and a milestone for the entire field of molecular biology. With the advent of more tags in subsequent years—negating the need for luck—the technique was eventually used to locate thousands of disease-causing genes throughout the human genome.

In 1993, Wexler and an international team of more than fifty researchers, nicknamed the Gene Hunters, identified the exact nucleotide sequence of the Huntington's disease gene. The news was greeted with spectacular admiration from across the world, draping the cover of the most famous newspapers and scientific journals. Twenty-five years after Wexler had learned that her mother had Huntington's disease, she could now recite the sequence of As, Ts, Gs, and Cs that caused her mother's early death. She could spell out the molecular source of her own fears.

Huntington's disease turned out to be an affliction of arithmetic. Toward the beginning of the culprit gene, the DNA sequence CAG appears several times in a row. In a normal person, the three nucleotides are repeated (CAGCAGCAG . . .) up to thirty-five times before moving on to the rest of the gene. In people who develop Huntington's disease, the three-nucleotide sequence repeats forty or more times in a row. To figure out who would get the disease, a laboratory technician simply had to count the number of CAG repeats in a person's Huntington's disease gene. Fewer than thirty-five repeats, and a person was safe. More than forty, and the sentence was sealed. In between was a gray area, an arena where only time would give the answer.

Wexler never did take the genetic test she helped to create. Her movements eventually revealed what she had always feared: she developed her mother's affliction. Today, her neck cranes, her fingers curl, and her legs dance in the unending motions of Huntington's disease. Her family's illness is on display wherever she goes.

But Wexler's work may soon form the basis of a treatment for her disease. Once scientists located the gene that causes Huntington's disease, researchers figured out that the abnormal DNA is not a problem in itself. Rather, the affliction develops because the body synthesizes dangerous proteins according to instructions from the faulty DNA. With more than forty CAG repeats in the Huntington's disease gene, the body creates a flimsy protein that clogs brain cells, producing symptoms of the disease.

To stop the condition from developing, researchers designed a drug that prevents the body from synthesizing proteins based on the directions coded in the Huntington's disease gene. The medication works like noise-canceling headphones; the disturbance—the abnormal DNA—still exists, but it may no longer cause a person to experience symptoms. If the drug is as effective as researchers hope, entire families will be released from the chokehold of their DNA. Huntington's disease will become preventable, or even treatable.

～～～～

THE PIECE OF PAPER mailed back from a laboratory revealed Amelia Ellman's genetic fate with just two digits: four and six. Amelia, the high-flying aerialist who had built her life around precision movements, had been walking around with forty-six CAG repeats in her Huntington's disease gene ever since she was born. Her body had been coded to self-destruct.*

But if the science works out as neurologists hope, Amelia will never actually develop Huntington's disease. She could raise children to adulthood. She might play with her grandchildren, her body still moving with graceful control.

For now, Amelia spends her time honing her muscles and her mind. She has purchased an exercise studio, where she teaches her students yoga and other movement methods. Her fingertips extend and her legs stretch in a carefully controlled display of muscular power. Outside the building, a purple sign with white letters hangs from the brick exterior: YOUR MOVEMENT HAVEN, the name reads. It is exactly the kind of shelter she has always wanted.

---

* Many patients ask how genetic diseases can develop in adulthood when the gene causing the condition has been present since birth. The answer remains unclear, but likely has to do with environmental factors, cellular aging, and the particular characteristics of the genetic change.

Chapter Two

LA BOBERA DE LA FAMILIA

HUNTINGTON'S DISEASE TURNS OUT TO BE unusual among causes of dementia, because it is a one-gene, one-disease affliction. Everyone with Huntington's disease has an abnormality in the same gene.

Alzheimer's disease, on the other hand, is almost always the result of numerous poorly understood genetic and environmental risk factors. Most people with Alzheimer's disease do not have a particular gene they can point to as the lone culprit for their ills. It is extraordinarily difficult to study such a heterogeneous population, so Alzheimer's disease researchers have looked for rare families who carry single genetic mutations that cause the condition. One such family, wracked by early-onset Alzheimer's disease for more than two centuries, has now become one of the most valuable research cohorts in the world.

The clan hails from Antioquia, Colombia, a region of sturdy people and rough terrain. The land is dotted with cows, horses, and soaring wax palms whose tops scrape the low-hanging clouds. Mountains turn to valleys and then back into peaks with such regularity that it is difficult to go anywhere without summiting or descending. For generations, few Colombians moved and few arrived. The isolation, perpetuated for generations, led to a hereditary catastrophe.

In 1984, Dr. Francisco Lopera was a second-year neurology resident in Medellín, the capital city of Antioquia. He was gregarious and easygoing, with fluffy eyebrows that overshadowed his cheeks like the peaks of Antioquia's mountains. Lopera had grown up in a town near Medellín and had dreamed of studying outer space and unidentified flying objects. Finding as an adult that earthly matters were just as interesting and more easily accessible, he had turned his attention to neurology.

With a reflex hammer and a tuning fork dangling from his white coat pocket, Lopera walked into his clinic one morning to find Hector Montoya sitting on the examination table, flanked by his children. Still in his forties, Hector had become so confused that he could no longer work. Farming tasks that he had done deftly for years now bewildered him. He would erupt into tears and keel over in laughter at odd times. Hallucinations distorted his reality. He had been smart and stable earlier in life, his children told Lopera, but now it was as if he had become a different person.

Lopera hospitalized Hector for intensive evaluation. Other than his cognition, Hector's neurological examination was normal. His reflexes pinged appropriately. He could stand up from a chair without using his arms. He could walk across the room with ease. But Hector could not identify his own age, the current date, or that he was in the hospital. He no longer recalled his children's names, or even how many offspring he had. Each morning, he would wake up unaware of how he had ended up in the hospital or whether he had ever met Lopera before.

Lopera ordered a scan of Hector's brain. Slipping the stiff black-and-white films onto a lightboard, he considered the size and shape of each of the lobes in the pictures. Quickly, he spotted an abnormality toward the center of the images: the structures that facilitate memory, normally plump and seahorse shaped, had nearly disappeared.

Lopera compiled a list of potential diagnoses, conditions common in people as young as Hector. He considered Huntington's disease, which had made international headlines the year before with Nancy Wexler's work. He thought also of frontotemporal dementia, a disease that caused victims to lose inhibitions and empathy. But Hector's symptoms and brain scan did not suggest either condition. After consulting with colleagues, Lopera made a diagnosis that would change his career: only in his mid-forties, Hector suffered from a disease of old age. He was plagued by early-onset Alzheimer's disease.

When most people conjure an image of a person with Alzheimer's disease, they imagine long-retired people with gray hair. The picture is often accurate. More than 80 percent of people with Alzheimer's disease are over age seventy-five. Lopera's own grandmother was the quintessential case: a woman who had lived long enough to meet her great-grandchildren but could not remember their names. Just 3 percent of people with Alzheimer's disease are younger than age sixty-five, and far fewer are under the age of fifty, as Hector was when he came to Lopera's clinic.

Questioning Hector's children, Lopera discovered that his patient was just one of many people in the family to suffer profound memory loss. Hector's father and grandfather had suffered the same symptoms, at roughly the same age. The story was common among other people in the community, too. The syndrome was so widespread that locals had created a name for it: *la bobera de la familia*, "the idiocy of the family." They used *la bobera* for short.

Theories on the origins of la bobera abounded. Some of Hector's relatives believed the condition was a curse inflicted by a priest whose parishioners had stolen from the collection box. Others said it was caused by touching a specific bark. Still others held even more ornate stories of witchcraft. Hector's children had barely heard of DNA, but they knew la bobera was hereditary. They could see it had infested their family tree.

Lopera soon found that his clinic could not contain his curiosity. By car, on horseback, or on foot, he traveled for hours on cracked roads and overgrown paths to interview people in Hector's extended family. He kept track of individuals on index cards, jotting notes about each person's symptoms and gathering information about who was related to whom. Back in his office in Medellín, he rearranged the cards as he discovered new familial connections, building an ever-growing family tree that soon included hundreds of people.

The enormity of the problem became clear only a few months later, when another patient with early-onset Alzheimer's disease made an appointment at Lopera's neurology clinic. The woman's father, uncle, grandfather, and great-grandfather had all endured unrelenting memory loss beginning in their thirties or forties. Lopera reconstructed the woman's family history, recording who was affected and who was not. Several generations back, the woman shared an ancestor with Hector

Montoya. The two occupied separate branches of the same gargantuan family tree.

Working with colleagues, Lopera pored over old ledgers from church parishes and notaries, searching for evidence of where and when la bobera had first appeared. The oversized record books, brittle and dusty, contained meticulous tables of births, deaths, and marriages going back over two hundred years. Some people had died of a condition described as "softening of the brain," which Lopera thought might be the same as la bobera. He traced the diagnosis back to a husband and wife of European descent who were born in Medellín in the mid-eighteenth century. The couple had at least three children. Over two hundred years later, their offspring numbered in the tens of thousands.

On a winter morning in 1992, Lopera sat in a large lecture hall, listening to Dr. Kenneth Kosik give a talk. Kosik was an intrepid American neurologist with expertise in the biology of Alzheimer's disease. He was soft-spoken and inquisitive, the type who could not get more than a few sentences into a conversation without showering the other person with questions.

After the lecture, Lopera approached Kosik to discuss the work he had been doing over the past ten years. "I have a family here that has early-onset Alzheimer's disease," Lopera said. Kosik was skeptical at first, figuring Lopera had in mind a family of three or four individuals. But as Lopera described the extent of his genealogy studies, Kosik leaned in. His fingers fidgeted. The Colombian spoke broken English, and the American knew only a few words in Spanish at the time, but the volley of questions and answers quickened as if the language barrier had collapsed. If Lopera was correct about the diagnosis, Kosik understood, the implications would be staggering. By the time the men parted for lunch, Lopera had invited Kosik to come to Medellín and meet Hector's family. Kosik canceled his flight back to the United States, and a lifelong research partnership was born.

As Kosik talked with Hector's family in the villages of Antioquia, the next step for the project became increasingly clear: Lopera and Kosik would need to prove la bobera was really Alzheimer's disease. Until then, Lopera had made the diagnosis based on cognitive testing and imaging alone, a method that was accurate only about 80 percent of the time. To prove their case, Lopera and Kosik would need to show that la bobera caused the same microscopic changes that Dr. Alois Alzheimer had discovered a century earlier. They would need to get ahold of a victim's brain.

〜〜〜〜

DR. ALOIS ALZHEIMER WAS a young and boisterous neurologist who counted among his accolades both a medical degree from a renowned institution and a citation for disturbing the peace. Among colleagues, he was known for his mastery of microscopy, a skill he honed while working on a doctoral thesis about the cellular life of earwax. In 1888, he took a job seeing patients at an asylum in Frankfurt, Germany. It was there, thirteen years later, that he came across a young woman who had lost her memory.

Little is recorded about the patient, Auguste Deter, before her arrival at the hospital. She had married a railway clerk in rural Germany in 1873 and had given birth to a daughter some years later. Photos show her with a lean face framed by straight hair that reaches a few inches beyond her collarbones.

At home, Auguste had begun to make mistakes in the kitchen when she was just in her mid-forties. She left out ingredients from long-memorized recipes, so that her cooking became unpredictable and undelectable. She would wander from room to room in her apartment, unable to find the bedroom or the living room. Increasingly paranoid, she hid valuables under furniture and behind books. Later, when she could not remember where she had put them, she would accuse people of stealing them. She screamed for hours in the middle of the night and lashed out at her husband, convinced he was having an affair with a neighbor.

Her husband despaired, taking Auguste to a doctor who promptly wrote out a prescription: "[Auguste] needs treatment from the local mental institution." Her husband went home, packed a suitcase of her clothing, and set out with her to the Asylum for the Insane and Epileptic in Frankfurt. She was just fifty-one years old. She would not leave the facility alive.

Dr. Alois Alzheimer had a habit of keeping detailed notes on patient encounters, so it is no surprise that he recorded a word-for-word account of his first conversation with Auguste on the day after she arrived at the asylum in November 1901:

"What is your name?"

"Auguste . . ."

"How long have you been here?"

"Three weeks."

"What do I have in my hand?"

"A cigar."

"Right. And what is that?"

"A pencil."

"Thank you. And that?"

"A steel nib pen."

"Right again."

Midday, Alzheimer returned with another set of objects. He showed each to Auguste, who named them with little difficulty. Minutes later, after the items were put away, she had no recollection of the interaction.

Alzheimer set a paper in front of his new patient and instructed, "Write 'Mrs. Auguste D.' " Auguste wrote "Mrs.," then forgot the assignment.

With more testing, Alzheimer gleaned that Auguste still remembered long-known facts. She could name the color of snow, soot, and the sky. She could calculate 6×8 and 9×7. She could recognize a spoon, a toothbrush, and a key by feeling each object in her hand with her eyes closed. If Alzheimer held up three fingers, she could count them correctly. Immediately afterward, she would have no idea how many fingers he had shown, or even that he had put up his hand at all.

Alzheimer examined Auguste several days in a row, each time expanding his understanding of what she could and could not do. Her symptoms reminded him of older patients whom he had seen with senile dementia, but Auguste Deter was not senile. She was barely fifty years old.

Alzheimer prescribed hours-long water baths for Auguste's confusion. He recommended sedatives for her agitation.

Despite the effort, she was soon restricted to an isolation room at night; otherwise, she would spend the dark hours creeping into other patients' beds, extracting shrieks of surprise that sounded through the corridors.

As months wore on, Auguste became convinced she was at home, welcoming guests. "My husband will be here shortly!" she would declare, although she no longer recalled his name. She spent most waking hours apologizing for not having her house ready, fretting about being unprepared for a dinner party that would never happen.

Alzheimer wrote a final entry in Auguste's chart in June of 1902: "Auguste D. continues to be hostile, screams, and lashes out when one wants to examine her. She also screams spontaneously, often for hours, so that she has to be kept in bed."

Auguste lived another four years, but Alzheimer never saw her alive again. The year 1901 had been a time of academic inspiration for him, but it had also been one of personal tragedy. His wife had suddenly become sick in early January. By late February, he had watched her coffin descend into a rectangular plot at the edge of a Frankfurt cemetery. Widowed at age thirty-seven, he became a single father of three children under the age of six. With little tying him to Frankfurt, he moved the family to Heidelberg and then to Munich, where he became a pupil of the pre-eminent psychiatrist Emil Kraepelin.

Just a hundred steps from the asylum at the University of Munich, Alzheimer rented a third-floor apartment in a classic late-nineteenth-century German building. The shrill sound of agitated patients emanated from the ward at all hours of the night, drifting easily across the short distance into his living room. It was as if he could never fully leave his new work, and he liked it that way; after his wife died, research offered an escape from mourning. His sister had taken over much of the work of raising his children, and he was free to immerse himself in the intricacies of the human mind.

Setting up his laboratory space at the University of Munich, Alzheimer positioned microscopes atop long tables next to a window. He set up stools that could be adjusted for people of different heights. He purchased a camera lucida, a light-bending tool that reflected the images from a microscope onto a flat surface where he could trace what he saw onto paper.

Students shuffled into and out of the laboratory each morning and afternoon. As the space filled with chatter, Alzheimer would walk pensively from desk to desk, teaching his pupils about the methodology of microscopy and the angles of anatomy. By the end of the day, his cigar was invariably left burning at the desk of one student or another, discarded unconsciously in the passion of a teaching moment.

As frenetic as Alzheimer's days became in Munich, he never forgot the case of Auguste Deter. He contacted his colleagues in Frankfurt every so often, asking for updates on her status and reminding them to call him when she died. Twice he intervened with his own financial support and academic clout to keep Auguste from being sent to a different hospital, where he might lose track of her.

On April 9, 1906, an intern from the asylum in Frankfurt phoned Alzheimer to report that Auguste's life had ended the previous day. At Alzheimer's request, the intern preserved and packaged her brain and sent it, along with her chart, to the laboratory in Munich.

Flipping through Auguste's thirty-one-page file, Alzheimer saw her admission note from the winter of 1901, then several of his own notes from subsequent months. After he had left the institution, Auguste had continued to deteriorate. "Completely stupefied," someone had written in her chart in 1905. "In a crouched up posture in bed. . . . Plucks with her hands at the bed cover," a clinician documented later the same year. For the last month of her life, nurses had given Auguste daily therapeutic baths, a treatment that probably exacerbated her bedsores. She stopped eating. Her weight plummeted to sixty-eight pounds. She developed a fever on a spring day, and a few mornings later she was dead.

The acrid smell of formaldehyde enveloped Alzheimer's laboratory as he unpacked Auguste's brain. Even at a glance, he could tell the specimen before him was atypical. It was just a fraction of the size it should have been.

The outside of a normal brain has hills and valleys, so that if you drag your fingers over the organ, your hand will bob up and down like a ship in calm seas. In Auguste's brain, the hills had become thin and flimsy. The valleys had grown wide.

Alzheimer sliced Auguste's brain into pieces, then soaked the samples in a dark solution. He washed the slices, heated them, then washed them again, before mounting each between two pieces of glass. Turning to the tables by the window, he set the slides under the lens of a microscope and brought the images into focus.

With Auguste's brain cells magnified, Alzheimer could

see why her mind had faltered. A long, tangly substance had accumulated in Auguste's neurons. Nearby, in the supportive tissue of her brain, dark plaques appeared like piles of seeds. Both findings reminded Alzheimer of specimens he had seen from older patients with senile dementia, but Auguste's case was far more severe. Her entire brain was riddled with the tangles and plaques.

Alzheimer had not given much thought to the structures when he had seen just a few of them in the brains of older patients, but now he wondered whether they might in fact be the cause of disease. Positioning the camera lucida so that the microscopic images were reflected onto a piece of paper, he traced dozens of neurons, each marred by stringy tangles and seedlike plaques. Eager to share the findings with fellow physicians, he collected the drawings and prepared a lecture for an upcoming conference in the nearby city of Tübingen.*

The response at the conference was hardly what he expected. "Doctor Alzheimer from Munich will now present 'On a Peculiar, Severe Disease Process of the Cerebral Cortex,' " a colleague had declared, motioning to him to ascend the podium.

Alzheimer began by describing Auguste's strange behaviors. He told of how she could no longer read or write, and how she substituted words with descriptions, like "milk pourer" for "cup." Finally, he showed drawings of the plaques and tangles he had seen in her brain, assuming his audience would marvel

---

* Coincidentally, the University of Tübingen is the same institution where Friedrich Miescher discovered DNA more than half a century earlier.

at the sheer burden of the strange structures. "Taken all in all," he said proudly, "we clearly have a distinct disease process before us."

Silence prevailed in the room when he finished speaking. A moderator asked the audience for questions, but nobody offered any. Alzheimer stared at the people seated in front of him, searching for an interested face. The room remained noiseless. "So then, respected colleague Alzheimer, thank you for your remarks," the chair of the meeting finally said. "Clearly there is no desire for discussion." Alzheimer took his seat in the auditorium. Later that day, the same audience would erupt into boisterous debate in reaction to a lecture on the cause of excessive masturbation.

Disappointed but not deterred, Alzheimer continued to toil at the laboratory bench, hoping to understand the role plaques and tangles played in dementia. In tandem with his basic research, he treated patients at the University of Munich, where he soon met more young people with symptoms similar to Auguste's. As he had done after her death, he examined each patient's brain postmortem. Just as he expected, the organs were full of plaques and tangles.

While Alzheimer continued soliciting interest in his discovery, it was Emil Kraepelin—the renowned psychiatrist who had brought Alzheimer to Munich in the first place—who finally attracted the world's attention. In writing a new edition of a textbook in 1909, Kraepelin devoted several paragraphs to Auguste's story. He described the plaques and tangles Alzheimer had seen in her brain, expounding on the question Alzheimer had proposed at the conference in Tübingen: Did Auguste's disease represent a severe form of senile dementia, or a new condition entirely? In the last paragraph of the section, Kraepelin referred to "Alzheimer's disease" for the first time in published history. Within a few years, the term was used worldwide.

Alzheimer enjoyed fame from his eponymous disease for less than a decade. In early 1915, a simple cold evolved into a corrosive infection of his heart. He tried to continue his research but found his body made this impossible. His kidneys failed, his lungs filled with fluid, and his mind lapsed into delirium. By the morning of December 19, 1915, death loomed. His children crowded around his bed and prepared to become orphans. Alzheimer took his final breaths at fifty-one years old—the same age Auguste had been when the two first met.

———～～～———

BY THE TIME FRANCISCO LOPERA and Kenneth Kosik were enmeshed in the mystery of la bobera, plaques and tangles had become critical to the diagnosis of Alzheimer's disease. To prove la bobera and Alzheimer's disease were the same condition, the two men would need to show that la bobera caused the same microscopic changes that Alois Alzheimer had noticed a century earlier in Auguste Deter's brain.

In 1995, Lopera called the children of a fifty-six-year-old woman who had just died of la bobera. He asked if he could have the woman's brain to examine in his laboratory. La bobera was probably Alzheimer's disease, he explained, but he needed to look at a victim's brain to prove it. He said the work might help doctors find a cure for the lethal affliction that now threatened each of the woman's children, too.

The family refused. A growing underground organ market had made people increasingly suspicious of anyone who asked for body parts, and Lopera's request was considered no different. He tried to reassure the family that his intentions were noble, but the woman's children were resolute.

So Lopera drove five hours to the woman's wake to talk with her family in person. He paid respects at the coffin laid out in the living room, then turned to the woman's next of kin. He said he had grown up nearby and was not an intruder taking advantage of their community's misfortune. Echoing what he had said on the phone, he reiterated the importance of microscopic studies for for finding the cause of la bobera.

As hours passed, all but one of the woman's children agreed to Lopera's request. The holdout, a former policeman with suspected ties to drug trafficking, left in anger and returned drunk, demanding twenty million pesos. Lopera said he could not pay the family for their donation; all he could do was promise that he had no intention of selling the matriarch's brain. Finally, the man capitulated. On the way to the church, Lopera and a pathologist stopped briefly at the local infirmary to remove the woman's brain before her funeral. The rest of her body was carried onward to the burial. In the meantime, Lopera soaked her brain in formaldehyde and prepared to send it to the United States.

Not long after, a colleague carried the brain aboard a commercial flight to Boston, Massachusetts. From there, the man took a taxi to Kosik's front door. "Here you go," the man said, handing over the brain in a box.

The next day, Kosik took the organ to his laboratory. He

sliced the tissue into thin pieces and soaked them in dyes that highlighted particular molecules. Under a microscope, the woman's brain had the plaques and tangles that had come to define Alzheimer's disease. With more than seventy cases of la bobera in an extended family that numbered in the thousands, Hector Montoya and his relatives were—and still are—the largest cohort of people with familial Alzheimer's disease ever found.

Lopera and Kosik understood that if families with la bobera were willing to participate in research trials, they could subvert the biggest barriers to finding a treatment for Alzheimer's disease. Scientists had spent decades enrolling patients in drug trials based on clinical diagnosis alone, without proving that participants had plaques and tangles. Twenty percent of people diagnosed this way turn out not to have Alzheimer's disease at all—so one in five people who participated in the trials did not even have the disease in question. Since most patients were over the age of sixty-five, even those who did have Alzheimer's disease often suffered from other conditions that also hindered their cognition, making it difficult to tease out which symptoms were specifically attributable to Alzheimer's disease. In contrast, la bobera is molecularly uniform. It always causes plaques and tangles. Since those who develop the condition are only in their forties, they rarely have other medical conditions that could account for their cognitive struggles. Compared to typical cases of Alzheimer's disease, la bobera is far less vulnerable to confounders. It is a purer malady.

Studying la bobera also avoided another source of statistical noise. Alzheimer's disease progression is highly variable, so that some patients experience a slow decline over more than a decade, while others free-fall into a rapid descent in just a few years. We cannot predict the rate of progression for most people with the disease, so it is difficult to prove that a potential treatment has blunted the expected course. Who is to say that a person who receives a medication and then declines slowly was not destined to have the same gradual slide even without the drug? Compared to typical cases of Alzheimer's disease, the course of la bobera is fairly predictable: children with the condition develop symptoms at roughly the same age at which their parents became sick. They decline similarly, with a pace in line with that of their parents. History repeats itself from one generation to the next, a fact that turns out to be terrible for victims but fortuitous for researchers.

TO PREDICT WHO IN Hector's extended family would develop la bobera, Kosik began looking for the gene that causes the disease. He considered performing the type of analysis that scientists had used to find the Huntington's disease gene, but Kosik worried that blood samples from people with la bobera had been compromised. Many of the tubes, which Lopera had been collecting since the 1980s, had been frozen and thawed an untold number of times as refrigerators in the war-torn country lost and regained power. Most of the DNA, Kosik found, was too degraded to use in experiments.

So instead, Kosik capitalized on genes that other scientists had already found to cause early-onset Alzheimer's disease. He contacted researchers at Washington University in St. Louis who were studying a gene named *presenilin 1*. Shipping samples from South America to the Midwest, he soon found exactly what he had been looking for: people with la bobera had a mutation in the *presenilin 1* gene.* Toward the middle of the gene, the nucleotide sequence was mutated from G*A*A to G*C*A. There was a typo.

Normally, the *presenilin 1* gene instructs our brain to build a protein that acts like a waste-processing center. The protein chops molecules into pieces, sending some fragments to the cellular trash heap and others to be recycled and used anew. With the Colombian family's mutation, the balance between discarding and recycling is thrown off-kilter. Molecules that would otherwise have been processed into reusable material are instead broken into toxic parts that accumulate in plaques—the exact microscopic hallmark Alois Alzheimer noticed in Auguste Deter's brain.

More than two decades after Hector first visited the neurology clinic in Medelin, Lopera and Kosik could finally attribute the centuries-long mystery of la bobera to a specific nucleotide in a particular gene. They could trace la bobera to a corrupted molecule. What's more, they could now identify who in Hector's family would go on to develop early-onset Alzheimer's disease, even before symptoms developed. The knowl-

---

* Before looking at *presenilin 1*, Kosik first looked at a gene on chromosome 21 that also causes early-onset Alzheimer's disease. The gene helps explain why people with Down's syndrome—who have an extra copy of chromosome 21—universally develop Alzheimer's disease if they live into their fifth or sixth decade.

edge would allow them to conduct one of the most unusual research projects ever organized.

Lopera had long wanted to run a clinical trial for Hector's extended family, but for years he had received the same response from drug companies: "How could we do clinical research in a country plagued by so much violence?" Lopera understood the concern; he himself had been kidnapped by guerrillas earlier in his career.* More recently, in 2000, he had paused his fieldwork for months after a colleague was kidnapped while trying to collect blood samples from people with la bobera. Colombia had long been one of the most dangerous countries in the world, and Antioquia was at the heart of the drug war that fueled everything.

But by the mid-2000s, Colombia began to emerge from decades of violence. Giants of the Medellín cartel were killed or incarcerated. Pablo Escobar died in a bloody shootout on a Medellín rooftop. Kidnappings declined. Seeing the improvement, scientists from an Alzheimer's disease research center in Phoenix, Arizona, reached out to Kosik and Lopera to see whether they wanted to pursue a clinical trial in Colombia.

In 2011, Lopera, Kosik, clinical-trial specialists, and representatives from some of the most promising drug companies in

---

* Years before, guerrillas had captured Lopera and taken him by horseback into the jungle, where they commanded him to treat a wounded comrade. After helping the man, Lopera was dropped off at his hospital with a torn five-peso bill and cryptic instructions to treat another guerrilla member who would come to the hospital in the future with the second half of the note. Lopera was told to document the man's gunshot wounds as accidental. In gratitude, the guerrillas left him a copy of *Das Kapital*, the foundational text by Karl Marx.

the world all converged on a generic-looking conference room in Phoenix. For days, the team reviewed candidate medications that could be used in a clinical trial. They sifted through data on drug mechanisms. They pored over results from studies conducted in test tubes and on animals. They mulled over Z-scores, T-values, and other statistical jargon.

Above all, Lopera and Kosik wanted to find a drug that would be safe for the people who took it. They planned to enroll participants who had not yet developed symptoms of Alzheimer's disease, in hopes that the medication would prevent memory loss from starting. Compared to studying people who were already symptomatic, the approach was more likely to succeed but was also more risky. Lopera had spent three decades building trust with Hector's family, and he knew the relationship might not recover if he endorsed a medication that hurt people who were still healthy. "These are normal people we are going to treat," he cautioned others at the meeting. "We cannot afford to do any harm."

The attendees finally settled on a medication that causes plaques to change from an insoluble clump to a soluble form that immune cells can mop up and dispose of. The medication, which would be given by infusion into a person's veins, is essentially soap for plaques. With $15 million from the National Institutes of Health, $15 million from philanthropists, and $70 million from the drug company that made the medication, the trial opened in 2013.

Results of the study will not be available until 2022, but families with la bobera have already changed the landscape of Alzheimer's disease. In the fall of 2019, a photo of Francisco

Lopera was printed on the front page of the *New York Times*. Underneath was the story of Aliria Rosa Piedrahita de Villegas, a Colombian woman who had turned seventy years old without any sign of dementia, even though her DNA contained the *presenilin 1* mutation that causes la bobera.

Piedrahita had become an outlier because she carried a second genetic mutation that hindered the development of tangles. While her brain was plastered with plaques, she had minimal tangles and few cognitive complaints by the time she succumbed to melanoma in 2020.* After her death, slices of her brain were sent to laboratories around the world, including to Kosik's office in the United States, where the samples are being used to uncover the molecular pathway that protected her brain from tangles.

The story of Piedrahita is a molecular exception to a molecular exception, an example so unusual that researchers have yet to find anyone else quite like her. "She has a secret in her biology," Lopera has said of her, adding, "This case is a big window to discover new approaches." Scientists are now searching for a medication that can recapitulate the effect of Piedrahita's mutation, preventing the buildup of tangles in the brain. The concept reinforces the idea that Lopera has been betting on for decades: that the cure for the common case of Alzheimer's disease will be found not in the general population but rather on the molecular fringe.

---

* Thanks to newly developed radioactive molecules that stick to plaques or tangles, scientists were able to prove while Piedrahita was still alive that her brain had significant plaques and only minimal tangles.

Chapter Three

HAS ANYONE SEEN
MY FATHER?

DANNY GOODMAN HAD ALWAYS BEEN WARM AND unflappable. As a young father, he would buy his children choc-olate milk before dinner, citing the importance of savoring life's sweetness. Family meals were boisterous with arguments about books and current events, the result of Danny's encouragement to read, to love knowledge, and to debate. His daughters were so enamored with his effusive character that they nicknamed him Fluffy, a moniker that stuck even as the girls grew into adulthood and Danny reached middle age. "Make mistakes," he told his children. "Own them. Learn something. Then move on and make different mistakes."

A born extrovert, Danny excelled in business. He bought nine-teen Burger Kings, then sold them and invested in thirty-seven Wendy's. He bought a Ground Round, a linen store, and an arts-and-crafts center. When a business magazine asked him in the 1990s if he was an entrepreneur, Danny had replied, "I think that's too glitzy a term. There isn't any magic to what I do."

In 2012, a prominent technology publication ran a story on Danny's newest venture, a website that sold one type of wine per day until the supply ran out. The article traced the unlikely success story of Danny and his brother, both in their sixties, neither computer savvy, who had built one of the most profitable online wine retailers in the country. Just a few years into its existence, the company had grown into a multimillion-dollar Goliath with tens of thousands of customers. At an age when many people were winding down their careers, Danny was only growing his.

"I need a CFO," he said to his son Russell, shortly after creating the wine company. "I'd like it to be you." Like his father, Russell had an animated face and a large personality, with a warmth and gregariousness that made him immediately likable. He had a PhD in molecular pathology, an MBA in finance, and a résumé filled with scientific accolades. Russell had not sold a bottle of wine in his life—he had never even run a direct-to-consumer business—but he knew he wanted to keep the nascent company in the family. Heeding his father's request, he quit his job in pharmaceutical research and went to work selling wine to the masses.

Danny began to change six months later. He stopped hugging Russell. He purchased hundreds of DVDs online, impulsively clicking the Buy button six or eight times, so that—to his wife's dismay—multiple copies of the same DVD would bulge out of their mailbox. He spent thousands of dollars on shoes and ate massive amounts of pizza, without concern for his shrinking bank balance or his expanding waistline.

At work, Danny looked at a February financial review and asked where the twenty-ninth and thirtieth days had gone. He misspelled *cabernet* on the company home page and peppered wine listings with exclamation points: *Pinot!!!Grigio!!,*

Me!!rlot!!!! He made last-minute changes to shipping plans, so the company was flooded with orders for bottles they could not supply. Whenever anyone called him on the phone, the shrill ring would hang in the air until calls went to voice mail; Danny no longer felt obligated to pick up the receiver.

Then one morning Russell found his father watching pornography on his computer at work. The screen faced the office door, so anyone who walked by could see the nude images. Employees blushed. Conversations stopped.

When Russell confronted him, Danny insisted he had done nothing wrong. "This is my business," he bellowed. "I can do whatever I want."

Overwhelmed by worsening snafus and sandwiched between his father's enterprise and his scorn, Russell created a dummy version of the company's website for his father to work on. Each morning, Danny would log on to his computer and find the familiar setup of buttons and text boxes. The system functioned exactly as his old interface had, except for one feature: unbeknownst to Danny, none of the changes he made were ever saved to the website's code. At age sixty-two, he had been covertly forced into retirement.

Everyone seemed to have a different opinion on what had caused Danny's behavior to change. A neurologist diagnosed him with attention deficit disorder and told him to see a therapist for depression. A family friend became convinced it was a stomach issue. Another was sure the symptoms were related to back pain.

Russell's PhD in biology guided him in a different direction; he suspected his father's problem might be genetic. Danny's maternal grandfather had developed dementia at a young age. A cousin on the same side had lost the ability to talk, and family lore was that an uncle had died of a rare type of dementia called Pick's disease. Russell wondered whether the three unusual stories might be linked by his family's DNA. He considered whether Danny might have inherited the same condition. With mounting unease, Russell realized the affliction could one day destroy him, too.

～～～～

THE CONDITION THAT STALKED the Goodman family was first described more than a century earlier by Arnold Pick, a stern German psychiatrist. Pick was a multilingual bookworm whose library grew like moss on every surface of his home. "My dear, if you continue in this way," his wife had once exclaimed, "I am going to move out with the children and you can have the apartment for yourself and your books."

Academically, Pick was raised by the kings of neuroanatomy. In 1874, he worked under Theodor Meynert, who identified a part of the brain that is critical for memory networks. Pick did research with Carl Wernicke, the foremost expert in the biology of language. He practiced medicine under Karl Westphal, who described a now-famous group of nerves that control eye movements. All three mentors had their names attached to important brain structures; all are memorialized in medical textbooks still used today. By the end of his career, Pick's name, too, would become synonymous with a major neurological entity.

But Pick's career was bounded by financial strain and

bureaucratic limits. After graduating from medical school in 1875, he accepted a position at a grim-looking asylum on Katerinska Street in downtown Prague.* The facility was dirty and dated, so overcrowded that in the evenings staff would spread mattresses on the floors between beds to accommodate the huge number of patients. Political tension and language problems strained interactions between German-speaking staff and Czech-speaking residents who saw their doctors both as their individual captors and their country's occupiers. "Das muss jetzt ertragen werden," Pick would say, frustrated with the need to waste time navigating politics. *This, too, must be endured.*

Pick eventually became an academically cautious professor who preferred a conservative interpretation of facts to the risk of scientific inaccuracy. He refused to write a textbook, worrying that something he described as factual might turn out to be incorrect. He believed there were some riddles neither science nor philosophy could answer. "Ignorabimus," he would say, quoting a well-known researcher with a similarly conservative mindset. *We will never know.*

Just as Auguste Deter prompted a seismic shift in Alois Alzheimer's work, a patient named Anna Jirinec became a catalyst for Pick's research. Anna was a seventy-five-year-old tailor's wife who came to Pick's clinic in 1900 after neighbors found her traipsing through their property. She had been

---

* Thanks to its near-mythical status as a house of chaos, the name *Katerinska* had become synonymous with pandemonium, much as the term *bedlam* was born out of conditions at Bethlem Royal Hospital in London.

literate and calm for most of her life, but her behavior had become bizarre over the previous three years. She had ripped up the vegetables from her garden, leaving a mess of seedless soil. Fleeing home for no obvious reason, she would wander aimlessly along dusty roads and over strangers' properties until someone recognized her and took her back to her husband. Relatives wondered if she had become deaf, since she no longer seemed to understand their scolding. Others believed she had gone insane.

Anna's speech was a hodgepodge of nonsensical phrases when she first talked with Pick. "Jesus Mary, no clothes there!" she said, angry that the asylum's staff had labeled her clothing and put it away in a cupboard. When Pick asked her age, she responded, "The poor thing is dead, little Anna is not dead."

As Pick studied Anna's use of language, he found something unusual: she had little understanding of words and objects, but other parts of her intelligence remained intact. She could communicate clearly with gestures. She remembered recent events, like where she had hidden a coin several days earlier. Most patients with dementia—like Alzheimer's patient Auguste Deter—had marked memory loss by the time they struggled with language. Anna, on the other hand, had profound language difficulties while her memory still worked well.

Four years later, Anna died at the same asylum where her husband had dropped her off. At Pick's request, her brain was removed and brought to his laboratory for examination. With the organ resting on the counter in front of him, he began to understand what had happened to her.

Our brains usually remain roughly symmetrical throughout life, but the left half of Anna's brain weighed 13 percent less than the right. Her left temporal lobe, which mediates language, had withered. The finding explained why she had so much trouble speaking and understanding people. She was not deaf, as her family had thought. Her ears worked fine. She had simply lost the ability to comprehend language.

Anna's case challenged a fundamental idea that Pick had long taken as fact. For decades, his mentors in neuropsychiatry had defined two categories of brain diseases: focal and diffuse.

Focal processes, like tumors and strokes, affected a particular part of the brain and spared the rest of the organ. The cause of injury would be obvious to the naked eye, and a victim's neurological examination would be normal, with the exception of signs that were directly related to the part of the brain that was damaged. The handicap, along with its structural cause, would be limited to a particular region of the brain.

Diffuse processes attacked the whole brain indiscriminately, causing the entire brain to shrink like an ice sculpture in a warm room. According to Pick's mentors, senile dementia was exclusively a diffuse process. It affected the entire brain at roughly the same rate, leaving no neuron out of reach.

Anna's case suggested the prevailing belief was wrong. Pick did not see evidence of a stroke or tumor that could account for the difference in weight between the left and right sides of her brain. Rather, he believed, senile dementia alone had caused her left temporal lobe to shrink while the rest of her brain remained relatively normal. Contrary to what he had been taught, Pick started to think that dementia was not always a diffuse disease. Rather, he suspected, it could be a focal affliction just like strokes and tumors.

Pick knew he needed to find more patients to convince others that Anna's case was not singular. Over fourteen years, waiting until he was confident he would not be proved wrong, he reported on more than a dozen patients like Anna, whose frontal or temporal lobes had shrunk more than the rest of their brains.* In each case, the patient's symptoms correlated with the area of the brain that was affected: those with small temporal lobes tended to have difficulty with language, and those with paltry frontal lobes struggled with disorganization and disinhibition.

Pick's work eventually gained the attention of Alois Alzheimer, who by then was renowned for discovering his eponymous disease.† More than a decade younger than Pick, Alzheimer was already a celebrity with immense academic and financial resources. Alzheimer had access to an enormous laboratory with state-of-the-art equipment; Pick's research space was a roughly hewn room with three crowded desks where students sat shoulder to shoulder. While Pick needed his salary to sup-

---

* Part of the reason these studies took so long was that there were no imaging techniques at the time to evaluate a living person's brain. Rather, Pick needed to wait for each patient to die before he could remove the person's brain and assess its geometry.

† Pick and Alzheimer were in fact scientific nemeses, in part because Pick had been a staunch opponent of Emil Kraepelin, the German psychiatrist who had mentored Alois Alzheimer.

port his family, Alzheimer had worked for free for many years, living off the inheritance from his dead wife. Pick could write case reports about curious syndromes, but only Alzheimer had the resources to pursue a methodical, microscopic study of the strange phenomena.

So it is no surprise that the person who uncovered the microscopic hallmarks of Pick's cases was not Arnold Pick but Alois Alzheimer. Alzheimer identified two patients with symptoms similar to Anna's. After the patients died, he brought their brains to his laboratory to look at their tiniest structures under a microscope—the same process he had used to identify plaques and tangles years earlier in Auguste Deter's brain.

Alzheimer thought that the patients might have accumulated plaques and tangles in a non-uniform manner. He suspected that areas of the brain that had shrunk would bear a larger burden of the abnormal structures than those that had maintained a normal size.

Instead, Alzheimer found something astonishing: the stringy tangles he had discovered a decade earlier in Auguste's brain were entirely absent from the sample in front of him. Likewise, there were none of the seedlike plaques that by then had become well-known markers of Alzheimer's disease.

In place of plaques and tangles, Alzheimer saw dark, oval spots inside the patients' neurons. "The question is," he wrote in 1911, "are these patients to be assigned to senile dementia?" Had Pick discovered a new version of an old disease, or a separate entity entirely?

By 1922, the condition Pick described became known as Pick's disease, etymologically erasing Alzheimer's contribution to the discovery.* As with his teachers, Pick's name was inked into the annals of neurology, attached to a novel type of dementia that was soon recognized as a separate condition from Alzheimer's disease.

A year later, Pick attended a performance of Beethoven's third Razumovsky Quartet. Overwhelmed by the sound of the strings, he turned to his daughter and declared, "Das ist zum Sterben schoen." *That is so beautiful, one might wish to die.* Pick was healthy at the time, but it was the last concert he ever attended. The following year, a stone clogged his urinary tract. When a violent storm cut off electricity to Prague's main hospital, a surgeon tried to remedy his affliction by candlelight. Pick suffered overwhelming infection over the following days. Doctors could do little to help him; it was many decades before antibiotics would be discovered. Fully aware of his impending death, he lost consciousness as his family and two of his students sat by his bedside.

～～～～

IN THE AFTERMATH OF Pick's death, the disease he defined became an object of perplexity. The dark, oval blobs Alzheimer had discovered under a microscope turned out to be rare in people who had the symptoms Pick described. Alzheimer's seminal case of Pick's disease had, unfortunately, dealt with an uncommon variant.

---

* It is not hard to guess why Alzheimer's contribution to the condition was scrubbed. The scientist who coined the term *Pick's disease* was a student of Arnold Pick.

Without an identifiable microscopic hallmark, research on Pick's disease fell into dispersed disarray for half a century. Suspected cases were described across the world with a myriad of laughably cumbersome names like "disinhibition–dementia–parkinsonism–amyotrophy complex," "rapidly progressive autosomal dominant parkinsonism and dementia with pallido-ponto-nigral degeneration," and "familial multiple system tauopathy with presenile dementia." With inconsistent nomenclature, nobody knew whether Pick had discovered a single disease or many.

It was only in the 1990s, using the technology Nancy Wexler's team had used to find the Huntington's disease gene, that researchers began to fully understand the genetic scope of the condition Pick had described. Working separately, eight research groups around the world studied families who suffered from syndromes similar to those Pick had seen in Prague. Slowly, every team came to the same conclusion: the disease under study—which by then had acquired the more concise name of "frontotemporal dementia"—was linked to a small section of DNA in the middle of chromosome 17. Every patient with the disease had a mutation in the same gene. Under a microscope, all of their brains had accumulated the same abnormal protein. For a fleeting moment, the field was unified.

Then, things crumbled. In 2006, researchers found a second gene on chromosome 17 that likewise caused the symptoms of frontotemporal dementia. The new gene, named *GRN*, led to buildup of a different protein from the one that scientists had discovered before. Further data would show that the symptoms of frontotemporal dementia were just as likely to be caused by one protein as the other. The result finally answered the question scientists had wondered about for decades: molecularly speaking, Pick had in fact discovered more than one disease entity.

It was around this time that Danny Goodman, the increasingly inappropriate founder of a multimillion-dollar wine company, finally agreed to see a neurologist who specialized in cognition. Looking at images of Danny's brain, the doctor quickly located the problem. Just as Anna Jirinec's brain had wilted in one area but not another, Danny's brain was under siege from a focal process that was slowly shrinking his frontal and temporal lobes. Hearing about his family history of peculiar cases of dementia, the doctor counseled that the symptoms had probably not come about at random. Rather, Danny's disease was likely the result of a DNA mutation.

The doctor was right. Danny Goodman's bloodline carries a mutation in the *GRN* gene on chromosome 17. The *GRN* gene is written in the same four-letter code as all human DNA: *A*s, *T*s, *G*s, and *C*s. In Danny's family, a single nucleotide within the gene is changed from a T to a C; the sequence reads CC*C*GG instead of CC*T*GG. It was this minuscule change, first occurring by chance in one of his ancestors, that now co-opted Danny's personality.

"Did you hear? I'm demented," Danny would blurt out after he was diagnosed with frontotemporal dementia. When his elderly father passed away, he walked around the memorial service asking, "Has anyone seen my father?" He ridiculed one of his grandchildren's classmates on social media. He solicited prostitutes, a behavior that nearly ended his marriage. Secrets once sequestered in nighttime whispers were suddenly displayed to the world without filter. Intimacy became impossible.

Friends did not understand that dementia could cause hypersexuality and loss of empathy. They thought it started with forgetting keys and appointments—the way the disease is portrayed in movies. Most friends withdrew. Overwhelmed, Danny's wife stopped talking with the few people who tried to stay close. With a single DNA mutation, she lost her partner and most of her community.

Then, as more of Danny disappeared, the situation became easier for his family. In the later years of frontotemporal dementia, apathy reigns. Victims are often content to sit in a chair for hours, staring at a wall. They no longer have any interest in arguing. In the end, frontotemporal dementia looks like any other dementia: a frail person lying in bed, muttering, picking at bed sheets.

"He's just a husk of a person," Danny's son Russell described toward the end of Danny's life. "He's bedridden. He doesn't speak at all." In the fall of 2018, Danny died of frontotemporal dementia.*

<hr>

* Patients often ask me what the immediate cause of death is in people with dementia. One of the consequences of dementia is that we lose the insight and automatic functions that previously protected our bodies. This can lead to infections, blood clots, falls, and loss of appetite, among other issues, which often become the proximal cause of death in those with dementia.

THE GOODMAN FAMILY'S ENCOUNTER with frontotemporal dementia is not over. As soon as Russell learned his father had a detectable genetic mutation, he wanted to know whether he had it, too. On a windy day in February of 2015, Russell sat across from a genetic counselor. He recounted the story of his father, his father's cousin, and his father's uncle. The counselor drew squares and circles—signifying men and women, respectively—and filled them in to indicate who had frontotemporal dementia. She asked Russell if he was sure about proceeding with the test; he had no symptoms of the disease, and there was no known medication that could stop the condition if he turned out to carry his father's mutation.

"It will not be a problem for me to know," Russell reassured the counselor. As much as he had grown into his new role at the wine company, he was still a scientist at heart. He had internalized his father's belief in the inherent value of knowledge. Hoping for good news, he rubbed two cotton swabs on the inside of his cheek and tucked them into an envelope to be sent to a laboratory.

Weeks later, he learned that he had not only inherited his father's creativity and appearance. He had also inherited his father's *GRN* mutation.

After hearing the news, Russell went to a bar with his wife and cried over hard liquor. Glassy eyed with a tonic of shock, disappointment, and fear, the couple took stock of the damage Danny's disease had done to their family. They wondered how they could drop off their children at school and go to work with a sense of catastrophe looming. Russell fell into depression, heavy with guilt as he imagined burdening his family.

But as is common for people who find out their genetic status, whether good or bad, Russell settled into a new steady state over the next several months. He began to think about frontotemporal dementia through a scientific lens, as if it were some other disease he had encountered during his career as a molecular pathologist. He still thought about his genetics on most days, but other things occupied his mind first: wine tastings, feeding the dog, his daughter's Bat Mitzvah. Like most people, he could endure chaos for only so long.

"This disease can end with you," he has told his three children, his voice imbued with both relief and resignation. For Russell, finding power over his family's affliction has been paramount to survival. He has encouraged his children to use in vitro fertilization when they want to have children of their own. Doctors would merge sperm and egg in a laboratory and check whether each of the resulting embryos carry the *GRN* mutation. Only embryos that do not have the mutation would be used to create offspring.* If his children take advantage of the

---

* Doctors created this technology in 1989 to help a couple whose first child suffered from cystic fibrosis, a lung disease that at the time carried a life expectancy of less than twenty years. If the couple conceived again without assistance, they would have had a 25 percent chance of having another baby with cystic fibrosis. To remove the risk of passing on the disease, which is caused by a genetic mutation, doctors performed a modified form of in vitro fertilization. They retrieved eggs from the woman's ovaries and mixed them with the man's sperm, creating embryos. Then the doctors did something

technology, they can have biological children without worrying about passing on the mutation that killed their grandfather. The disease can be extricated from their family line.

Scientists may soon have a treatment to offer Russell, too. Doctors believe that people with *GRN* mutations develop frontotemporal dementia because their cells create less of a critical protein than normal cells do. To remedy this, scientists have created drugs that either increase production of the important protein, or protect the protein from destruction. The approaches are akin to increasing capital by either printing more money or putting a hold on spending. The medications are still under study in humans, so we do not yet know if they work. But Russell Goodman is holding out hope.

---

that had never been done before: they pulled away a single cell from each embryo and determined the DNA sequence in the cell's cystic fibrosis gene. Knowing that embryos with the feared mutation would grow into babies with cystic fibrosis, the researchers selected embryos without the mutation and placed them in the woman's uterus. One of the embryos grew into a fetus, and nine months later, the woman gave birth to a healthy baby girl who did not carry the mutation that causes cystic fibrosis. The procedure became known as "preimplantation genetic testing," referring to the fact that DNA sequencing happened before the embryo was implanted in a woman's uterus.

Part Two

REBELLIOUS PROTEINS

PROTEINS ARE MOST FAMILIAR IN EVERYDAY life as something we eat. If you open your pantry, you will see the protein content of breakfast cereals, tuna fish, and virtually every other food item printed on a nutrition-fact label on the side of its packaging. When we eat protein, our body breaks it down into components and reconstructs the parts into new proteins that keep our cells running. We build these proteins to help us survive—and we would not be alive without them—but they can also put our brains on the fritz.

The history of proteins begins almost a hundred years before the discovery of DNA, in a Parisian laboratory where a chemist named Antoine François de Fourcroy gingerly treated parts of dead animals with heat and acids while the French Revolution raged outside his window. Fourcroy was a counselor to Napoleon and one of the most brazen brain researchers of his era, known for exhuming coffins to harvest human organs for his work.

It was Fourcroy's animal experiments that ultimately led him to discover proteins. Over several years, he learned to extract three substances from animal remains: albumin, fibrin, and gelatin. Albumin was a firm, white material that appeared when he heated egg whites; fibrin was a tough substance that emerged from clotted blood and animal muscles; and gelatin was a mush that formed when he boiled animal tendons.* Each of the three products came from a different animal part, but Fourcroy found they all shared something in common: they all contained nitrogen.

Few scientists recognized the accomplishment at the time, probably because it was difficult to appreciate small scientific discoveries in a world where muskets, guillotines, and revolution dominated everyday life. Fourcroy moved on to other endeavors, like establishing the metric system and expanding the periodic table. After a whirlwind career in science and politics, he dropped dead at age forty-five, his demise rumored to have been hastened by the shame of not receiving a promotion he had wanted.

The early threads of protein science were largely left to nothingness for half a century, until a young Dutch researcher named Gerardus Mulder picked up where Fourcroy left off. Mulder was a burgeoning food scientist who would eventually become a celebrity among beer aficionados for uncovering the chemical secrets of brewing. In turning to Fourcroy's old papers, Mulder found something far more universal. He discovered a substance that linked all living things.

Following Fourcroy's protocols, Mulder produced his own samples of albumin, fibrin, and gelatin. He also learned

---

* You have probably encountered albumin numerous times without realizing it. When you make a sunny-side-up egg, it is albumen that turns the outside of the egg white.

to extract a similar substance from wheat, so the materials at his disposal now included derivatives from both animals and plants. To expand on Fourcroy's work, Mulder determined which atoms, other than nitrogen, were in the four different samples. He expected to find that each was made of a unique combination of atoms. After all, he reasoned, each had come from a separate source.

Instead, Mulder found the opposite. Albumin, fibrin, gelatin, and the wheat extract—each culled from a completely different origin—were all made from nitrogen, oxygen, hydrogen, and carbon atoms, in roughly the same ratio. At a molecular level, the extracts were essentially identical. Somehow, Mulder realized, mammals, birds, and crops had all come to contain the same molecule.

Excited but bewildered, Mulder wrote a letter to his mentor, Jacob Berzelius, describing what he had found. Berzelius was a revered figure in modern chemistry; he had discovered two elements and invented the still-used shorthand for describing chemical substances. Reading of Mulder's results, Berzelius was quick to join in amazement. If the findings were correct, he realized, there had to be a single substance at the root of all living things. He replied to Mulder with a letter that would change the field of biology, for the first time giving a name to the molecule that existed across so many organisms. Berzelius proposed the name "protein," explaining that he had derived the term from the Greek word *prōteios*, meaning "in the lead," since the substance Mulder had found was surely the "original or primitive substance."

Taking Mulder's data a step further, Berzelius proposed

a model for how such molecular continuity could have come into existence. Proteins, he suggested, originated in plants that were eaten by herbivores, which were eaten by carnivores, thereby linking all plant and animal life with a single, shared molecule. The idea was not exactly correct, but its essence has been revived in modern times; in 2017, scientists posited that proteins, rather than DNA-like molecules, might have seeded early life on earth.

In the decades after Mulder published his results, scientists began to understand that proteins were in fact giant molecules created from small building blocks called amino acids that are strung together and folded into three-dimensional shapes. At its core, every amino acid contains nitrogen, oxygen, hydrogen, and carbon atoms in an identical arrangement; this was the common chemical content Mulder discovered in the 1800s. Branching off this identical backbone are wildly different "side chains" that define each individual amino acid. The arrangement is similar to a charm bracelet, where each piece contains an identical link attached to a variable structure.

Scientists have likened proteins to "nature's robots." The molecules can cut, merge, and transform other molecules. They can provide the structural support both to maintain a cell's shape and to mediate the magnificent, dynamic process of splitting a cell into two. They dictate where fingers grow on a developing fetus, and they detect when the time comes for uterine contractions. Even the texture of hair—whether it curls into ringlets or lies straight on the head—is determined by proteins. The mass of proteins in our cells is so gigantic that if you could remove all the water from a human body, you would be left with a pile of molecules almost half of which would be proteins.

Proteins are multitalented and numerous, but they can also be tempestuous molecules. They can rebel. A protein meant to transport vitamins can instead settle into the walls of the heart, causing the organ to become so stiff that it can no longer contract forcefully enough to circulate blood around the body. The same protein can cause the liver to swell and the kidneys to stop making urine.

Self-harming proteins are a particularly daunting threat when it comes to the brain. The cast of mind-altering proteins most often inflicts terror in one of two ways. First, proteins that are meant to defend us from infection can instead mount a coup d'état, attacking our own brain cells. Second, proteins built to keep our cells running normally can fold into toxic shapes that prevent our neurons from firing correctly.

Unlike DNA mutations, rebellious proteins tend to exact a rapid toll on cognition, often bringing us from normalcy to near death in just a few weeks or months. Since the maladies are not encoded in DNA, we usually cannot predict who will develop disease until symptoms begin. For most of history, there was nothing doctors could do at that point. Physicians merely watched as their patients' brains were swallowed by cantankerous proteins.

Today, the outcome is far better for many of these conditions. For the first time in history, we are starting to meet people who stared down the barrel of a rebellious protein and survived to tell their story. Here are some of them.

Chapter Four

A ZOMBIE APOCALYPSE

TWO DECADES BEFORE SHE BECAME SICK, LAUREN Kane was born to parents in the final stretches of a volatile separation. Her father moved out of their house in the middle of an argument and never returned to see her learn to walk or talk. On the verge of homelessness, her mother worked two jobs to pay for an apartment for her three children and herself. A half sister became a de facto parent, and Lauren spent most evenings playing by herself in a Fisher-Price castle that took up much of the living room.

Tiptoeing near poverty, Lauren grew up to be an academic force. In German class, she absorbed new words and grammar as if they were part of her native tongue. She gained acceptance to a renowned summer program for exceptionally bright students. At her high school graduation, she crossed the stage wearing the salutatorian stole.

In college, Lauren began writing fiction about characters with atypical family structures. Her laptop was crowded with stories of adopted children, single parents, stepparents, and found parents. "It's actually pretty common," she would say about upbringings like hers, "but nobody's writing about it."

By the time she took her clothing and comforter home, diploma in hand, she wanted only to return to school to write more.

Everything changed one August morning in 2016. Lauren woke up early and went downstairs, where her mother was gathering items from the refrigerator to make breakfast. Clutching mugs of coffee, the women chatted as sausage and eggs crackled in the frying pan. They moved to the barstools at their tall kitchen table and ate breakfast as they lamented the recent death of their cat. Swooning over old pet photos, neither woman was aware that it would be the last normal conversation they would have with each other for months.

After eating, Lauren returned to her bedroom and fell asleep. "What's for breakfast?" she asked her mother an hour later.

"We already ate," her mother replied. She noted the strangeness of her daughter's comment but thought little of it; time had become irrelevant without the didactics and deadlines of college courses. Lauren had been engrossed in *The Walking Dead*, a postapocalyptic zombie horror show that she watched episode after episode, like chain-smoking.

Lauren went to sleep again, then woke up midday. "What's for breakfast?" she asked.

By early evening, her forehead was warm and her steps were unsteady. She stumbled in the carpeted hallway. Descending the stairs, she gripped the banister so tightly that her fingertips turned white. Frightened, her mother helped her to the car and drove her to the hospital.

Sitting in the emergency room, Lauren recognized she was surrounded by nurses and doctors but could not remember why they had come. "Mom, what happened that we had to take you to the hospital?" she asked, not realizing that she was the one on the gurney. "I'm losing time," she repeated every few minutes, as if the statement was novel each time. "Do you think this could be a virus in my head?" she asked her mother.

The curtain opened and a scrubs-clad doctor came in. "What year is it?" he asked. Lauren answered correctly.

"What state do you live in?"

"Pennsylvania."

"Can you count backward from one hundred?"

Suddenly, as if occupied by a spirit, Lauren reached for the doctor's chest and gripped his shirt. She thrust him across the room, then dug her fingernails into the arm of a startled nurse. Her mother moved to calm her, but Lauren pushed her to the floor. The beat of footsteps echoed across the emergency department as nine security guards rushed like a tidal wave toward the room. They burst in loudly, each draping himself on one part of Lauren's body or another in an effort to control her.

"Don't you see it, she's a walker," Lauren yelled, pointing at one of the guards. A page rang out overhead, calling for reinforcements.

"Is she on PCP?" one guard yelled to Lauren's mother, who was still struggling to stand up from the floor.

"Oh my god," said another guard, placing Lauren's words into their Hollywood context. "She thinks she's in *The Walking Dead*."

～～～

Subdued with sedatives, Lauren was soon admitted to the hospital's neurology ward, where she became increasingly unpredictable. At times, she seethed with aggression. At other moments, she was calm but confused. She drifted in and out of the world of *The Walking Dead*, mistaking hospital staff, friends, and family for characters from the show. Occasionally, there were moments of lucidity. She would remember that her cat had died and start to cry. She would recognize her mother and talk of feeling worried about her. But within a few hours, the confusion would invariably set in again.

Days passed with little progress toward making a diagnosis. Lauren's doctors tested her for seizures, strokes, and infections. Everything was negative. According to blood tests and brain scans, she was normal. It was only in talking with her that it became apparent something was wrong.

Lauren's mother began recording their conversations, hoping the audio files might one day yield a diagnosis. She captured the snippets on her phone and gave the collection the bland title "Lauren Hospital Audio 2016." In one clip, which she labeled "Feeding Time; Zombie Apocalypse," she captured her efforts to nourish Lauren:

"Want some more melon?" she asked.

"Look for supplies, or look for families or friends we used to know," Lauren said. She spoke quickly, like a film being played at double speed.

"You mean because of the zombie apocalypse?" Lauren's mother asked, having already learned that it was better to meet her daughter in the fictional world than to try to convince her that she had lost hold of reality.

"Yeah," Lauren confirmed.

"OK, well, let's have some fruit first."

"I can hear them."

"Who are they?"

"OK, well, it was nice to meet you, Rick," Lauren said after a delay, calling her mother by the name of a character in *The Walking Dead*. "I guess I'm just going to run out and try to shoot shit. I've got to go because I got some walkers that are stuck to my arm. Nice seeing you."

"It was nice seeing you, too."

"There's so many of them attached to me now. It's hard for me to move."

"So many of what—the walkers?" her mother asked. "They're kind of restricting your movement?"

"Yeah." A rustling ensued.

"Where are you going?"

"I'm pushing it out of the way. I'll try not to lose my supplies." Lauren paused. "Apparently my legs are tied to the—" her voice trailed off. Then, with more curiosity than fright, she remarked, "That's so weird."

As Lauren's hospitalization stretched into a second week, her mother adopted the signature behavior of parents of children with unsolved diseases. When a doctor walked into the room each morning, she would be ready with a notebook of lined paper. She would jot down bits of what the physician said, asking for the spelling of unruly words like "e-n-c-e-p-h-a-l-i-t-i-s"—inflammation of the brain—and "l-e-u-k-o-c-y-t-o-s-i-s," a flood of white cells. She would underline words she thought were important, making sure to look them up when the doctor left.

As Lauren's world morphed into that of *The Walking Dead*,

her mother's world narrowed to Lauren's nine-foot-by-nine-foot hospital room. A plasticky chair that had been crammed into the space became her official place in the world.* Day and night, she kept watch over her daughter, hoping one of the always answerless doctors would burst into the room and announce the reason for Lauren's illness.

The dramatic moment never happened. Increasingly discouraged, Lauren's mother began looking for a diagnosis on the internet. Soon, she found an article about diseases that happen when the body creates a protein that attacks the brain. According to the article, there were several culprit proteins, each of which caused a particular constellation of symptoms. Some caused seizures. Others caused stiffness. One of the proteins—the first of its kind to be discovered—caused young women to have sudden bouts of psychosis. "That's exactly it," she thought, bookmarking the page to show the doctor the next day.

~~~~~~~~

BEFORE LAUREN'S MOTHER FOUND the article, a guard in the emergency room had come closest to diagnosing the cause of her ills. As Lauren fought off imaginary zombies, the guard—who had seen plenty of patients who were high on a variety of illicit drugs—asked if Lauren had used PCP

---

* At one point while Lauren was nearing the end of her hospitalization, I went into her room to replace her intravenous catheter. "No need to get up," I told her mother, whom I had obviously awakened. Her mother cared little for my offer. By the time I was set up to place the line, she was standing by Lauren's side, holding her hand.

(phenylcyclohexyl). Lauren had not, but her body had created a protein that had the same effect.

PCP was synthesized in the 1950s to solve a surgical problem. Until then, the only way to sedate patients for most surgeries was to administer general anesthesia. Doctors would wheel patients into the operating room and induce a coma with anesthetics that had the unfortunate side effect of stopping the natural drive to breathe. Quickly, a doctor would thread a hollow tube down a patient's throat and begin delivering bursts of oxygen while a surgeon performed the operation—removing an appendix, fixing a hernia, or something else. Then, if everything went as planned, the sedating medications would wear off, a doctor would remove the breathing tube, and the patient would once again breathe independently.

Anesthesia was a minor nuisance for the young and healthy but could cause problems for the elderly, obese, and sickly—the people most likely to require surgery in the first place. Some patients took days to breathe on their own after sedatives wore off. Others became dependent on the breathing tube for life. Still others died on the operating table as doctors failed to pinch-hit for the human heart-and-lung system. Sometimes sedation was so risky that patients were deemed to be ineligible for surgery, not because the cutting and sewing was dangerous in itself, but because the anesthesia was likely to kill them.

In the 1950s, chemists at a drug company in Detroit, Michigan, began looking for a solution. They collected molecules they thought could be useful for sedating people, then set out to modify each compound and observe the effects. They added and subtracted carbon, hydrogen, and oxygen atoms.

They stirred ingredients together, dried them under a vacuum, heated them, cooled them, and filtered them. Finally, they infused the modified substances into laboratory mice, cats, hamsters, dogs, and fish.

One of the molecules they synthesized, named PCP, caused an effect scientists had never observed before: it sedated the animals long enough to perform surgery, but it did not stop them from breathing on their own. After receiving an intravenous dose of PCP, the animals would fall unconscious. Scientists could then transfer pieces of skin, break bones, and even remove a stomach or gallbladder, all while their subjects continued to breathe on their own. When the anesthetic wore off, the animals would wake up and go about their usual laboratory activities.

Seeing what PCP could do, one scientist at the company called it "the most unique compound he had ever examined." With PCP, the scientists imagined, breathing tubes might become obsolete.

Triumphantly, the company marketed PCP as Sernyl, to evoke the serenity patients would feel while surgeons did their work. After small trials in humans, the drug gained approval from the Federal Drug Administration (FDA) in 1963. Shipments of the white powder made their way to hospitals across the country. Then—much to the company's disappointment—tales of startling side effects made their way right back.

Within months of the drug's release, stories of Sernyl-induced hallucinations proliferated. Some patients felt they were floating in outer space without arms or legs. A third of patients experienced confusion and agitation as the drug wore off. Doctors described "maniacal excitement" as patients became "noisy and abusive." People who had no predilection for violence became physically threatening to hospital staff and family members. While the best anesthetics wore off quickly, PCP stayed in the body for up to two days, yielding a trippy and prolonged recovery from operations. The purported miracle drug soon became one of the biggest failures in anesthetic history, and Sernyl became illegal in 1965—a mere two years after it received approval from the FDA.

PCP intoxicates by causing the limbic system, which processes emotions, to function independently from perceptions of the outside world. It causes people to dissociate. Feelings that would normally be kept in check by reality instead become a part of it. Aggression bursts forth, unmitigated, from deep inside the brain. Fragmented memories, often scary and strange, feel convincingly real. At the same time, information about the outside world is muted. Neurons meant to warn the brain when a body part is in pain instead fall quiet, sometimes permitting violent acts of self-mutilation.

At a molecular level, PCP disables a protein that floats at the outer edges of our neurons. The protein—called the NMDA receptor—has a remarkable job: it senses molecules released by neighboring neurons, determines whether the molecules are relevant, and then either opens or closes a tunnel between the neuron and the rest of the brain. In the process, the receptor helps us to gather information from our surroundings, to sort out what is salient, and to learn from experience.

When the tunnel in the NMDA receptor is open, connections between our neurons flourish and allow us to grasp new information. We become teachable. But the tunnel is a volatile structure. Open too long, and we can suffer relentless seizures, the product of neurons that are too interconnected for our own good. Open too little, and we can find ourselves in the grip of amnesia, dissociated from reality, or even comatose.

PCP works by sticking to the NMDA receptor and preventing the tunnel from opening. This is precisely why Sernyl was such an effective anesthetic; patients who received it stopped paying attention to the operating room around them. They could no longer learn that a scalpel was painful. They could not internalize the cutting and tugging of flesh. In the meantime, unbeknownst to anesthesiologists, the patients had been at the mercy of whatever reality sprouted from their minds—however violent and explosive that alternative experience was. "The profundity of PCP," writes Marc Lewis, an ex-drug addict–cum–neuroscientist, "lies in the embellishment of the self." With PCP, users inhabit only their own minds. Nothing else seems relevant.*

Lauren Kane had not used PCP, but her body had created a protein—called an antibody—that caused the same effect. The antibody stuck to her NMDA receptors and prevented the tunnels from opening. As a result, she was launched into the same dissociated reality that PCP users experience, with one notable

---

* Thanks to its unpredictable effects, PCP has been the least popular major drug of abuse in the United States for the past decade. Psilocybin (magic mushrooms) and lysergic acid diethylamide (LSD) have enjoyed a resurgence since the early 2000s, but PCP use has only continued to fall. In 2017, the number of new marijuana users in the United States outnumbered new PCP users by three million. With abuse so low, some hospitals have even taken PCP off of first-line drug tests.

exception: the effect of PCP wears off within a few days, but Lauren's intoxication lasted for months. With her body chronically producing a supply of the offending antibodies, it was, molecularly speaking, as if she were hooked up to an intravenous drip of PCP.

~~~~~~~

Our immune systems have been honed for millions of years to prevent the condition that took hold of Lauren. The average person makes ten billion distinct antibodies, each acting like an assassin with a particular molecular target. We make the antibodies prophylactically throughout our lives, before we are aware of what threats we will encounter. Lying in wait, the antibodies circulate in our blood and rest in our lymph nodes, each patrolling for a foreign molecule that fits the profile of its target. Most antibodies will go unused; we do not come into contact with an invading molecule that sticks to them. But some antibodies will encounter a target that attaches to them perfectly, like a puzzle piece finally reunited with its neighbor. When a match is made between antibody and invader, the body sounds a cellular alarm, calling immune cells into action.

There is one ground rule that is essential for this process to happen: an antibody that targets a molecule native to the human body must be obliterated. We want antibodies to fight the enemy but leave our own cells alone. Since antibodies are produced by a random process, those that bind to native human molecules inevitably arise from time to time in all of us. In most cases—through mechanisms still incompletely understood—our bodies get rid of these self-targeting antibodies without our even noticing.

But sometimes this quality-control measure fails. Antibodies that put our own cells in the crosshairs are allowed to proliferate. When they arrive in the brain, catastrophe ensues.

Then, the situation can get even worse. When an antibody recognizes its target, our immune system scurries into action, creating new antibodies that are more and more potent. In fighting infections or receiving a vaccine, our ability to develop better antibodies is beneficial, since repeated exposure to a germ or an inoculation causes us to create even-more-effective antibodies. But when it comes to autoimmune diseases like the one that afflicted Lauren, the pathway becomes a point of vulnerability. It allows the body to launch an ever more targeted assault on the brain.

This is precisely what happened to Lauren. At some point in the months before she became sick, she had unknowingly developed a small tumor on her right ovary. The tumor contained many different types of cells, including neuron-like cells bearing NMDA receptors. Mistaking the receptors for something harmful, Lauren's immune system produced millions of antibodies that stuck to the receptors and marked them for destruction. As the antibodies circulated in her brain, her neurons began to gobble up their own NMDA receptors, leaving fewer of them for her mind to survive on.

Instead of protecting Lauren, antibodies had caused her downfall. Scenes from *The Walking Dead* and thoughts of her cat's death—both memories from recent months—emerged in a distorted form and hijacked her reality. Fluorescent lights, white-coated doctors, and incessant recordings of blood pressures all failed to convince her that she was in the hospital.

Lauren's mother, who had found a story about NMDA receptor antibodies on the internet, asked her daughter's doctors to check for an autoimmune etiology for her illness. The request went unfulfilled. "It was as if they were saying, 'You're a mother, go away,' " she later recalled.

The rejection only made her more persistent. "If you won't do it, send us someplace that will," she said to the doctors. Bereft of diagnoses, the physicians acquiesced.

An ambulance took Lauren to a larger hospital, where results soon confirmed that her mother was right. Lauren's brain was under siege by an antibody that made her destroy her own NMDA receptors. An ultrasound showed the source of her ills, exposing the ovarian tumor that had probably been growing for months.*

As soon as the abnormal test result appeared in red numbers on Lauren's electronic chart, treatments began. Doctors prescribed high-dose steroids to quiet her immune system. Another medication helped her destroy the cells that made antibodies. A third medication served as a decoy, distracting her white blood cells so they spent less energy attacking her brain. Finally, surgeons addressed the origin of her ills; they took her to the operating room and removed the tumor from her ovary.

---

* The link between ovarian tumors and NMDA receptor antibodies had been established several years earlier. Today, female patients with antibodies against the NMDA receptor are routinely screened for ovarian tumors, which are found in about 50 percent of cases.

Days later, Lauren became reacquainted with reality. "What the hell is going on here?" she asked her mother. She looked at a whiteboard on the wall, where nurses had written and erased the date for weeks without her noticing. Now, she read the month in alarm. It was October. She had no memory of the prior two months.

~~~~~~~

THERE WERE SETBACKS ON the way to recovery. Even after treatment, Lauren's heart rate became unstable, at times soaring dangerously high, and at other moments falling so low that doctors worried the beating would stop altogether. She still had moments of confusion, partly because it took time to rid her body of toxic antibodies, but also because long stays in the hospital can themselves inflict fluctuating cycles of disorientation. She had not been outside in weeks. Aside from the small window in her room, there were few cues for when day turned to night and then back again.

In December of 2016, three months after her mother took her to the local emergency room, Lauren left the hospital for a rehabilitation center. Having spent so much time in bed, her muscles had weakened. She needed to do physical therapy to relearn how to walk. She had to develop tricks to keep facts planted in her mind. Her memory had improved, but it was not yet normal. It would happen at some point, doctors said—the quick-witted, academically savvy Lauren would return—but it would take months of medications and years of practice. What's more, her disease could come back at any moment. One in five people with antibodies against NMDA receptors will suffer a relapse.

In 2020, one of Lauren's short stories was selected for a national fiction contest—her first paid publication. She was considering writing a book. "I still don't have any memories of that time," she recalled of her hospitalization. "But I'm actually relieved about that. Based on things I've heard, I'm much happier not remembering any of it."

In any other decade, before the molecular cause of her disease was discovered, Lauren would have been admitted to a psychiatric ward at the start of her symptoms. Doctors would have remarked that she was precisely the right age for a first psychotic break. They would have sent her home with prescriptions for antipsychotics, instructing her mother to retrieve the pills from a local pharmacy and to never miss a dose. Drugs would have failed to cure Lauren's symptoms, since they would not have treated the underlying cause of her disease. Antibodies against her NMDA receptors would have continued to surge in her brain, dominating her personality and erasing a woman who had accomplished so much with so little.

Scientists now know that Lauren's disease is one of several conditions in which the human immune system makes antibodies that attack the brain. The list of similar afflictions grows every year as new syndromes are identified and new antibodies are discovered in laboratories across the world. People who would previously have been untreatable—and even undiagnosable—have now become curable. Their minds and lives are saved.

One of these people is Mike Bellows.

Chapter Five

THE MUSCLE MAN

MIKE BELLOWS MET AMY HOLMES IN THE EARLY 1980s, amid the clang of lockers in their high school hallway. Mike was brawny and handsome, a motorcycle-riding, church-going freshman born into a family of boilermakers who taught him that sweat was sustenance and endurance was paramount. Amy was tall and popular, two years older than him and adorned with a blond ponytail that bounced above the heads of other students.

Exchanging eye contact in the commotion between classes, the pair began a silent courtship as backpack-laden peers shuffled by. Amy would stare at Mike across the cafeteria, hoping he would finally discern her interest. Mike would make small talk with Amy at parties, searching for signs that she was attracted to him.

School years dragged onward and summers passed. Flirtation failed to evolve into something more. The teenagers swooned in silence, neither believing the interest was requited. Amy graduated high school and moved away. Mike followed two years later.

Mike served in the marines for four years, then became a

boilermaker like his father and grandfather. Dangling three hundred feet in the air, he wrapped hundred-pound chains around thousand-pound chunks of steel in nuclear power plants and oil refineries. He worked on weekends and early mornings, sometimes melting and molding metal for sixty days straight and then spending days off in the weight room at the gym. He married a woman he loved and had two children, then found himself signing divorce papers. Amy, meanwhile, was also raising children while navigating a separation.

Two decades after high school, Mike saw Amy at the general store in their hometown. He regaled her with memories of teenage times, quickly revealing his early love for her. Before the pair parted, he did what he wished he had done in high school: he asked her on a date.

Dinners rolled one into another. Mike made jokes about their age difference, which was now inconsequential but had seemed so intimidating to him years earlier. Amy found that Mike was the most versatile man she knew, as comfortable at a bar as at a black-tie event. The couple eased into the activities of new love, exercising together, cooking side by side, and introducing their children to each other.

Mike bought tickets to Turks and Caicos to celebrate Amy's fortieth birthday. He picked out an engagement ring from a jeweler near his home and tucked the box away in a closet to bring on the trip. He imagined dropping his knee to the white-sand beach, the crash of waves enveloping Amy and him in a soundproof intimacy.

Then, in the weeks before their flight took off, life began to splinter.

It started on a beach vacation in late summer, at an annual gathering that Mike had attended with longtime friends for years. He had always been difficult to ruffle and easy to amuse, but Amy noticed something changed that weekend: Mike avoided eye contact during conversations and instead stared into his beer bottle or out toward the ocean. When his friends planted their beach chairs and umbrellas together in the sand, he moved his to the other end of the shoreline.

Drenching night sweats began later the same month. Amy would reach her hand over to Mike's side of the bed in the morning and find him marinating. She set up fans, tried new detergent, and bought cooling sheets sewn from expensive fabric. Nothing helped.

Mike started smoking cigarettes again, a habit he had given up years earlier. Rather than the casual drags of high school, he inhaled desperately, as if enough nicotine could rescue him from mounting agitation. Even when he was alone, his heart seemed to contract with unusual force and spectacular speed, as if he were in a perpetual state of alert. He looked for a release—in exercise, in alcohol, in anything. Nothing worked.

As the weather cooled and the leaves began to turn bright colors, Mike and Amy boarded a plane for Turks and Caicos. Each hoped the warm breeze of a luxurious resort would return them to the easygoing relationship they had enjoyed before. On the flight to the island, Amy touched Mike's knee. His leg straightened suddenly, forcefully, kicking the underside of the seat in front of him. A stranger peered backward and scowled. Amy drew her hand into her lap.

Descending the stairs by the check-in desk at their resort,

Mike stumbled and fell onto the tiles. He seemed stiffer and easier to startle than before, as if the nimble strength that made him a talented boilermaker had left him.

Rolling her suitcase across the threshold of their hotel room, Amy pointed to the king-sized bed wrapped in white sheets. "Relax with me?" she said.

Mike's face softened. "I can lie down with you," he said.

Moments later, he threw his feet to the floor and stood up, agitated. He paced around the hotel room with heavy steps. His chest heaved. He punched his fist through the wall, sending chunks of plaster to the floor. Dust floated in the air.

Mike walked to the balcony doors and opened them, emerging onto the small landing. From the bed, Amy could see the open space below them, a distance large enough that even a man as sturdy as Mike would shatter on impact if he fell. She hoped he might freeze where he was, without going closer to the railing or nearer to her. She begged him to calm down. As seconds slinked by, she reached for the hotel phone and called security. Uniformed staff arrived quickly and worked quietly, coaxing Mike into a separate hotel room, where he spent the night.

Amy slept in their original room, debating in the early morning whether to leave the island and return home or to try to repair things with Mike. The first option felt unbearably sad. The second felt uncomfortably scary. She reached her arm across the side of the bed where Mike was meant to sleep, unsure whether she was relieved or disappointed to find his place empty and the sheets dry.

The following morning, the disillusioned vacationers met for breakfast at the hotel restaurant. "Are you OK?" Mike asked from across a white tablecloth. "I don't remember what happened last night." He thought of the engagement ring, still sitting in the darkness of a square box.

"You got really angry," Amy said, "and I had to call security."

"I don't remember anything. I'm sorry."

"I don't recognize you lately."

"I don't know what's going on, either."

The conversation happened calmly at first, with each person reiterating their hope for a return to their old relationship. Then Mike inadvertently knocked over a glass of orange juice. "Jesus Christ," he yelled, jumping backward and upending his chair. The dining room seemed to freeze mid-service as waiters and guests stared at his raging figure. He flung his hands in the air and swore, then yelled at Amy. He stomped off.

The pair stayed in Turks and Caicos until their planned flight home, but neither enjoyed the pristine beaches or the dry, hot weather. Mike wanted to calm down but could not. Amy wavered between anger and concern. Both wondered whether the story of their relationship would not end in the way they had come to imagine, culminating in the joy of reclaimed love and second chances.

When the couple returned home, Mike's behavior became even stranger. One day, Amy found him on his motorcycle in the front yard, staring at a fence. He told her he was stopped at a tollbooth and motioned as if he were holding out money. Then, as she watched, he fell over onto the grass. A day later, Mike thought Amy was a spider caught in a web outside his car window. "Are you OK?" he texted her. "It must be scary out there." He started to believe he was president of the United States, ordering people on missions that did not exist. Then Amy found him one evening sitting at the kitchen table wearing only underwear and boots, holding a fishing rod. "Let's get out of here," he said. "We have to get out." Amy called an ambulance.

"Isn't that my friend in the hallway?" Mike asked Amy the next afternoon when she visited him in his hospital room. She turned toward the doorway, where doctors and nurses walked by. "Is he on a wagon? A dune buggy?" he asked.

Mike's muscles began to cramp and spasm, set off by the smallest sound or the lightest breeze. The stiffness that began in Turks and Caicos exploded into episodes of whole-body contractions, as if Mike were being electrocuted. His jaws clenched shut, pushing his teeth like knives into his tongue, nearly severing the muscle from his body. His cheeks lifted forcefully toward his eyes, baring his gums in a horrifying facial contortion. Doctors tried hot compresses, cold packs, physical therapy, and muscle relaxants. The contractions continued, unmitigated, like firecrackers.

Mike's medical team moved through lists of diagnoses, sometimes settling on one disease or another for a short time before concluding that the story did not entirely fit. No single condition seemed to account for the way Mike's body and mind were locked in a constant state of tension. For months, he was moved in and out of hospitals and rehabilitation facilities, each time with more tests and still no answers.

Then, half a year after the sickness began, an ambulance happened to take him to the hospital where Dr. Greg McCarthy worked. McCarthy had a deadpan sense of humor and a wiry frame that skulked through hospital hallways. He was a lexicon of rare neurological afflictions, a Sherlock Holmes for the oddest of cases. For him, Mike's story had a familiar pattern: the spasms, the anxiety, the sweating. The constellation of complaints suggested a diagnosis.

~~~~~~~~

The history of Mike's condition dates back far before Mike himself, to a time when researchers were only beginning to understand how the nervous system works. It was the late 1800s—the era of Alois Alzheimer and Arnold Pick—and a young Spanish scientist named Santiago Ramón y Cajal had just discovered something remarkable.

Ramón y Cajal was an unlikely candidate to revolutionize the nascent field of neurobiology. The son of a domineering anatomist, he had little regard for rules and often found himself at the thin end of a whip. By age ten, he was thrown into a local jail for fashioning a homemade cannon and promptly blowing up his neighbor's property. After being forced into apprenticeships in haircutting and shoemaking—work he survived only by daydreaming of being elsewhere—he began accompanying his father to cemeteries to collect human bones for research. In the darkness, surrounded by tombstones, Ramón y Cajal fell in love with anatomy. Years of formal training later, he became a professor in Barcelona. There, he stumbled on a curious finding.

Until then, most scientists thought the nervous system was a single, continuous unit. The idea made practical sense, since physical continuity was the most obvious way that nerves in the brain could control fingertips and toes that were a long distance away. It was only logical that everything was connected. There was even microscopic evidence for the idea, since detailed images of nerves had not shown a definitive break between one nerve fiber and the next. Experiments suggested that the rest of the body was made of individual cells, but scientists thought the nervous system was an exception, that it was constructed of a single, mesh-like unit.

It was not. In 1888, Ramón y Cajal sliced up a collection of bird brains and soaked the pieces in a silver-laden solution that dyed nerve fibers black. Under a microscope, he found that the samples at hand did not contain the endless net-like structure he had expected. Instead, the cells had very clear termination points. There was contiguity, but not continuity.

Suspecting that the human nervous system might also contain individual cells, Ramón y Cajal turned his microscope on brain sections from his own species. He peered into the apparatus and saw the same phenomenon he had noticed in the avian samples. The human nervous system, like the rest of the body, was made of distinct units. It was composed of individual neurons.

Enthused, Ramón y Cajal created copies of his data to share with other scientists. The printing job was so costly that he had to give up hiring a nanny for his five children that year. Convinced that others would recognize his accomplishment, he mailed the monographs to researchers across the world, hoping for a deluge of praise.

Instead, the international community remained largely indifferent to the packages he had sent. Seeing the paltry reaction, Ramón y Cajal became entrenched in self-doubt, wondering if colleagues saw him as a fake. He agonized over his choice to write the articles in Spanish, a language he now realized his audience did not generally understand. Finally, he decided that the problem was distance. If he wanted to convince scientists that a central tenet of neurobiology was wrong, he would need to talk to them in person. He would have to force them to look at the evidence with their own eyes.

In 1889, Ramón y Cajal packed his microscope and slides and set out for a conference in Berlin. He barely attended the lectures on offer, and instead set up a display where he could show others what he had found in his own laboratory. On a table, he positioned his microscope along with a few others that the organizers had lent to him. Under each lens, he set up a slide of silver-dyed brain tissue.

The exhibit at first attracted a handful of skeptics, but the reaction soon turned positive. More and more people filed up to see the tiny images. By the end of the conference, the most renowned professors of neuroanatomy were swooning. The entire world—or at least everyone who was academically important—seemed to be interested in his idea. Over subsequent years, the audience from Berlin returned to their own laboratories to recapitulate Ramón y Cajal's work and confirm that it was correct. With very few exceptions, they found it was. By 1906, Ramón y Cajal had won a Nobel Prize.

Before long, scientists began to see that the new model of the nervous system presented a fundamental challenge. If nerves were built from individual cells, there had to be a way for the cells to communicate. They had to be able to talk. In some experiments, researchers noticed that neurons were nearly touching each other, separated by just a few nanometers that an electrical signal could easily traverse. In other instances, there was a perplexing forty-nanometer gap between the end of a neuron and the start of the cell it was communicating with. Scientists knew there had to be a way for neurons to send messages across the void, but nobody knew exactly how it could be done.

Then a German neurologist named Otto Loewi came up with a solution. Loewi, like Ramón y Cajal, was an unexpected champion of scientific breakthrough. Drawn to art since childhood, he had been pushed into the study of medicine by his father. By the end of his doctoral training, he was in remedial courses because he had skipped his assigned classes on the human body in order to attend lectures on art history.

Years later, Loewi would recount that the answer to the neuron problem came to him in a dream on a March evening in 1921. Grasping for a pen in the middle of the night, he jotted down his idea on a scrap of paper before pulling the covers back over his body and falling asleep again. When the sun came up, he looked at the note and found he could not read his own handwriting. The revelation had been lost.

The thought came to him again around three o'clock the next morning. Determined not to forget it, Loewi dressed and made his way to his laboratory in the dawn light. There, within forty-eight hours, he made the most important discovery of his career.

Loewi had long suspected that neurons could communicate by emitting and sensing messenger molecules, but until then he had been unable to show that such molecules were real. The middle-of-the-night idea that would make him famous was a simple experiment that finally proved he had been right all along. Working feverishly as the spring sun rose, Loewi isolated two beating frog hearts, one that was still attached to a nerve and a second that was not. He put the first heart into a container of fluid, then stimulated the attached nerve until it caused the frog's heart to beat more slowly. Then, he took the liquid from the chamber and used it to fill a separate container that held the second heart. Soaking in the fluid from the first heart, the second heart likewise began to beat less frequently. A molecule in the soup from the first heart had altered the behavior of the second heart. Just as Loewi had expected, the signaling molecules were real. Like Ramón y Cajal, Loewi went on to earn a Nobel Prize for his discovery.

Over subsequent years, the molecules emitted by neurons became known as neurotransmitters. These molecules, which include adrenaline, dopamine, and serotonin, have now become some of the most famous constituents of the human body. They are the subject of thousands of scientific papers and a number of best-selling books. Their names have even entered the vernacular.

The smallest neurotransmitter, a molecule named glycine, turns out to be critical to our ability to relax. Glycine allows us to release the tension in our shoulders and rest our heads on a pillow at the end of the day. For our neurons, sensing a burst of glycine is akin to drinking chamomile tea and taking an Ativan.

Like all neurotransmitters, glycine exerts its effect by attaching to a receptor molecule on the outside of our cells. It is at this step, this critical moment of detection, that things went terribly wrong for Mike Bellows.

~~~~~~

For months in 2016, Mike's body spasmed as if he had been infected with tetanus or poisoned by strychnine.* The convulsions became so powerful that his wrists and ankles had to be tied to the railings of his hospital bed so that he did not launch himself onto the floor. His face clenched in dramatic contractions, yielding a look of excruciating pain that horrified his family. Worried the spasms would block his airway, doctors put a breathing tube in his throat.

When Dr. Greg McCarthy saw what had happened, he suspected Mike's immune system had created an antibody that blocked his glycine receptors. To prove he was right, McCarthy needed to show that the culprit antibodies were present in Mike's spinal fluid. So one afternoon in February, a physician sedated Mike and rolled him onto his side. The doctor unwrapped a six-inch needle and pressed the metal bevel through the skin on Mike's lower back. The point of the needle passed one ligament after another, coming to a stop in the liquid that bathed his brain and spinal cord. A few millimeters of clear fluid dripped into a tube that doctors transported across campus to McCarthy's laboratory. Amid pipettes and petri dishes, McCarthy detected exactly what he had expected: an antibody that stuck to glycine receptors.

After six months, countless doctors, and hundreds of tests, Mike finally had a diagnosis that made sense. His own body

---

* Both tetanus toxin and strychnine cause striking, whole-body contractions by affecting glycine signaling. Tetanus prevents neurons from emitting glycine, whereas strychnine stops glycine from attaching to its receptor.

had attacked his nerves. He had become numb to the effects of a critical neurotransmitter.

Like Lauren Kane, Mike received steroids to quiet his immune system. Doctors filtered antibodies out of his blood. They administered a medication to destroy the cells that had created the antibodies in the first place. Unlike Lauren, Mike did not have a hidden cancer. There was no way to surgically excise the underlying cause of his ills. Nobody knew why his body had started to produce the offending antibodies. Research on his condition is ongoing in laboratories around the world, but we still do not know what triggers it.

Within a week of receiving treatment, Mike began to improve. The spasms stopped. His mind started to clear. "No matter what, I'm not going to stop loving you," he wrote one day in black marker on a small dry-erase board. He showed the words to Amy.

Within days, he regained control of his lungs. Doctors modified his breathing tube to allow him to talk again. "Amy, it's Mike," he said in a voice message. "I can talk. Call me at this number."

Soon, doctors removed the breathing tube entirely. For the first time in months, Mike spoke on his own. His jaws and esophagus relaxed enough for him to eat. He stood up from bed by himself and relearned to brush his teeth. He practiced enunciating different sounds: "ma-ta-ga" and "pa-la-ca," using his lips, teeth, and throat to form each syllable. Weeks later, he was discharged to a rehabilitation facility. Then, after more time working with physical, occupational, and speech therapists, he moved to a ground-level bedroom in Amy's house.

Mike began planning another proposal. Still not able to drive, he hobbled to the jewelry store where he had bought a ring months earlier, before the trip to Turks and Caicos. Figuring the old ring was bad luck, he returned it and chose a new one. With the modest backdrop of Amy's kitchen in the background, he unsteadily knelt down on one knee. "I've loved you since we were in high school," he said. "Will you marry me?"

Chapter Six

DEADLY LAUGHTER

THE DISEASE THAT STRUCK MIKE BELLOWS WAS caused by antibodies that should never have been made in any significant quantity. The proteins were harmful from the moment they were synthesized. The fact that they became numerous was, in itself, evidence that something was amiss in Mike's immune system.

But there is a second way the human body can wield a protein to sabotage the brain. Instead of producing a protein that is inherently harmful, the body can turn a normal protein into one that attacks the mind. The problem is not in the creation of a dangerous protein, but in the weaponization of one that is present even in normal circumstances.

The tale of these allies-turned-assassins begins with a bizarre, now-extinct disease called kuru. Today, kuru has changed our understanding of Alzheimer's disease, Parkinson's disease, and other common neurodegenerative afflictions. Its researchers have garnered two Nobel Prizes and international attention. But the origins of kuru lie in the remote highlands of Papua New Guinea, among tribespeople who almost never set foot outside their home territory.

Papua New Guinea takes up the eastern portion of a large, bird-shaped island whose beak faces Indonesia and whose feet point to Australia. It is the most rural country in the world, a place where sorcery and witchcraft remain participants in everyday life and tribal warfare still manifests in spats between neighbors. A tall mountain range cuts from east to west along the extent of the island like a spine. It is there, among the jutting peaks and lush valleys, that an unsuspecting public health officer named Dr. Vincent Zigas first witnessed kuru.

En route to Papua New Guinea in 1950, Zigas sat on an oil-stained sack in a small plane that rose and plummeted with every gust. The engine roared. Zigas fidgeted, taking inventory of the items in the fuselage around him: there were tractor parts, kerosene drums, and three sheep—future sources of wool and milk. Aside from the pilot, who would stay at their destination only long enough to refuel, Zigas was the sole human on the aircraft.

He had been eager to move to Papua New Guinea. Born in Estonia and educated throughout Europe, Zigas had resided in Germany most recently and did not want to weather the Cold War there. Signing up for the Australian public health service, he took a short training course before setting out on the assignment. As the plane prepared to land, he could not help but feel excited for the possibility of living in raw nature, among people less driven by consumerism.

Settling into a ramshackle cottage just before his thirtieth birthday, Zigas began seeing patients at the local hospital. The facility was bare-bones and run-down. Years of tropical grime had built up along cracks in its walls, so that the building had taken on the appearance of an organic, living structure. Inside, patients were seen by hospital staff who had little formal health-care training. Symptoms often took the place of diagnoses, and the few records that were kept proclaimed conclusions like "bellyache" or "fever."

Zigas had been told he would need to practice medicine only, but he soon realized he would have to double as a surgeon; there was nobody else in the region who could operate. Early surgical failures invaded his nightmares. A woman came in with a loop of bowel that had been caught in a hernia for five days. Zigas attempted to remove the strangulated tissue, but the woman's abdomen was already infected. She died in agony the following day. Another time, villagers brought a five-year-old boy to the hospital with burns covering most of his body. Zigas rushed the whimpering boy to the operating room, where he cut away charred flesh and flooded the child with fluids. When the boy's heart stopped, Zigas pounded on the small chest to restart the organ, but it never beat again.

Gradually, Zigas grew comfortable treating the prevailing ailments. He learned to repair wounds inflicted by wild boars, which were revered by local tribespeople but also had a tendency to gore them. He could extract teeth and fix bones after patients suffered farming and hunting accidents. He began to use maggots to clean wounds, an economical practice he adopted after a patient's leg accumulated a collection of the insects and underwent a surprisingly quick course of healing.

Zigas settled into life outside the hospital, too. He collected friends and learned to communicate in the local language. He found his way around the marketplace. But as integrated as he became in the world around him, it was still half a decade before he stumbled on the existence of kuru. The disease was so hidden in rural terrain that even people in Papua New Guinea had rarely heard of it.

At a party one evening in 1955, Zigas met a young drunkard whom the Australian government had sent to Papua New Guinea to build roads and schools. The man regaled Zigas with an account of a young woman he had seen weeks earlier in a distant village. The woman had sat next to a communal fire, her arms and legs shaking while her head tossed from side to side. Her movements had been so unpredictable, and her balance so compromised, that the Australian had worried she might roll into the flames. Local tribesmen told him the girl would succumb to sorcery within a few weeks. She suffered, they said, from "kuru."

Zigas was captivated. Whether kuru was a neurological affliction, a psychiatric condition, or something else, the disease was of obvious interest to him as the region's physician. "I can send a chieftain to take you there," the Australian offered.

A few months later, a written invitation arrived. Wrapped in breadfruit leaves fastened with jungle vines, the Australian's note was delivered to Zigas by a muscular man who identified himself as Apekono. "Follow Apekono," the letter said. "Grog and penicillin appreciated."

Zigas and his new guide set out by foot for the Fore tribal territories. Apekono walked ahead of Zigas, slashing branches that blocked their path and hollering when there was uneven ground ahead. Even with assistance, the territory was hostile. Leeches punctured the men's skin when they trekked through water. Mosquitoes swarmed them on land. At night, a moth larger than a man's palm, the aptly named *Casinocera hercules*, flitted around their lamplights. Under their feet, Papuan black snakes threatened to kill. The snakes were called *aguma*, meaning "to bite again," because of their tendency to repeatedly envenomate their victims.

After two days of hiking, the pair came to a cluster of mud-walled huts with thatched roofs. At one end, an unkempt path led to a crumbling structure. Apekono pointed to the doorway, where a young woman sat on the ground in the corner. Her face was entranced, her eyes staring past the men. Her body shivered as if from a cold wind, but the air was quiet and warm. When the woman tried to stand, a violent tremor shook her. She fell to the ground and giggled. Zigas recognized the symptoms as part of the disease described by the Australian drunk. The woman, Apekono confirmed, suffered from kuru.

Days later, they came to another village. Corpulent pigs loitered between the huts. One pig lay in a puddle next to a woman and a girl who together attempted to delouse it. Lice were a common snack for Fore tribespeople, but the woman and girl were too shaky to feed themselves. They flung the bugs upward unsteadily, opening their mouths in an unsuccessful attempt to catch the insects. The bugs arced through the air, landed on the ground, and scurried away through the mud. The woman pushed herself up, then toppled down, trembling. The girl, who appeared no older than ten, wobbled to a standing position and braced herself on a pole. Turning an empty stare in Zigas's direction, she displayed the rotting remains of neglected teeth. She laughed.

Zigas returned home a few days later, haunted by the images of kuru victims. He scoured textbooks looking for a similar disease, but he found nothing reminiscent of the giddy, shivering victims he had seen with Apekono. He wrote letters to colleagues in Papua New Guinea and around the world, pleading for someone to recognize the symptoms of kuru and make a connection to a more common affliction. The responses were all the same. Kuru seemed to be a singular disease.

~~~~~~

WHEN ZIGAS AND APEKONO reunited for a second expedition a year later, Zigas asked about the women and children they had seen on the prior trip. "They're all dead," Apekono replied, "but there are many more like them."

Apekono took Zigas to meet Taka, a childhood friend, who was collecting sweet potatoes when the pair arrived. Like Apekono, Taka had broad shoulders and muscular legs. But while Apekono's face was taut and youthful, Taka's skin was full of wrinkles. His skull housed anxious eyes.

The men held each other's scrotums—a traditional greeting gesture in parts of Papua New Guinea—and Taka set down his bag of potatoes. Apekono joked that his old friend, in digging up vegetables, had taken on a woman's job. Taka flushed, explaining that he was now the only one in the household capable of planting and harvesting crops. One daughter had been married off some years earlier, he said, and another had died of kuru as a teenager. His wife had recently developed the disease, too, after giving birth to their son. He pointed to the corner of the garden, where his wife crouched on the ground, shivering. She braced her hands on her knees and turned her body toward Apekono and Zigas, grinning with the signature vacant look Zigas now knew well.

An infant cried nearby. Taka fetched the baby and presented it to his trembling wife, who smothered it with unsteady affection. He tossed a few pieces of sugar cane and a dollop of pig fat into his own mouth, then transferred the masticated food to the infant; his wife was too malnourished to produce milk. Suddenly, Taka's voice boomed with anguish as he pointed at his wife and turned toward Zigas. "This woman cannot die," he said to the doctor, as if enough supplication could cause fate to yield in a kinder direction.

Zigas returned home a few days later and set to work writing more letters to colleagues, begging for help in figuring out what caused kuru. He invited visitors from around the world to come to Papua New Guinea and witness the disease.

Wondering if epidemiology would suggest an etiology, he sent emissaries into the Fore territory to collect information on kuru's origins. Elder members of the tribe could recount the history of the disease. It had first appeared in the early 1900s in the northwest corner of their area, before sweeping eastward, southward, and then northward again. Despite having no known infectious cause, kuru had grown to epidemic proportions in just a few generations. Over the same time, the disease exhibited two intriguing properties: first, it struck women and children in far larger numbers than men, leaving much of the male population spouseless; second, the disease ravaged the Fore tribe but rarely affected neighboring groups.

Hoping the mystery could be solved ethnographically, Zigas collected information on local flora, fauna, marriage rituals, and diets. He planned a third trip to the Fore territories, this time with government funding. He packed a Land Rover with food and medical supplies and bought a refrigerator to keep samples cold. The day before the expedition was set to depart, a visitor arrived unannounced.

Dr. Daniel Carleton Gajdusek was an American scientist who had heard of kuru from a Papuan health official. Clad in worn-out clothing and tattered sneakers, Gajdusek's dogged commitment to scientific proof had left him uninterested in his physical appearance. He had gained a reputation as a brilliant researcher and a defiant coworker, always hell-bent on scientific exploits, with little regard for the people around him. In Papua New Guinea, he looked for a fresh start.

At first, Zigas questioned Gajdusek's motives. He was unsure whether the American appreciated the human aspect of kuru, whether he understood that real lives were at stake. Gajdusek had never seen the excruciating reality of kuru firsthand. He had not looked into Taka's eyes. But Zigas also knew that he could not cure kuru alone. Gajdusek was as smart as any scientist he could hope to have on the project. So Zigas invited him to join the upcoming expedition, and the two set out together for the Fore tribal territories.

The newly minted duo set up their living quarters in a meager structure at the center of the kuru epidemic. Inside the building, a wooden table functioned as the laboratory, the examination space, the morgue, and the dinner table. Hoping a microscopic look at kuru would reveal the cause of the disease, the pair asked local tribespeople to donate the brains of people who had died of it. Many obliged, hoping the scientists could save their tribe from extinction. They carried the corpses of wives, sisters, and children across harsh territory to reach the makeshift laboratory, where Zigas and Gajdusek would preserve the organs and conduct their evaluations. In a photo of the setup from 1957, a kuru-infected brain spills over the edge of a small metal bowl in the middle of the table, next to two bottles of wine. Gajdusek peers into a microscope on the far end of the table as Zigas, clad in a T-shirt and shorts, takes notes. Neither wears gloves.

Kuru perplexed the men. They looked at red blood cells and white blood cells from people afflicted with the disease. They ran tests on urine and spinal fluid. The epidemiology of kuru suggested the condition was infectious, but they could not find evidence of a bacterium, virus, parasite, or fungus in the brain, blood, or urine of people who had died of the disease. They could not figure out what was causing the infection.

Hoping for help from across the ocean, the scientists mailed samples of affected brains to research laboratories around the world. Months later, the first clue arrived from the United States. A professor at the National Institutes of Health in Washington, DC, had examined the specimens under a microscope. Instead of the plump, well-defined neurons found in normal brains, those from kuru patients had a ragged, wilting appearance. Smaller cells, normally present in restricted numbers, had proliferated in place of dying neurons. The only thing kuru looked like, the professor noted, was a rare condition named Creutzfeldt-Jakob disease—the same disease Joe Holloway would be diagnosed with more than half a century later.

EVEN AS A MAN in his seventies, Joe Holloway had no chronic health issues. He did not take any medications, and he rarely sought the help of doctors. He was a retired chemist and a calculating tinkerer who could fix anything in his house without hiring a repairman. When the weather was warm, he would don gardening gloves and a hat, kneeling for hours over the immaculately kept planting beds in his backyard. He had an elegant wife, two children, and a growing cadre of grandchildren, all people he thought about each morning as he sat on a gray upholstered chair next to sliding patio doors, filling in the daily sudoku puzzle from the local newspaper.

In early 2016, Joe began losing his balance. He noticed the problem when he stood up from bed on the morning of January 4, then again the following day, and the next. He had always been sturdy and stable, but now his legs wobbled and his arms shook. Brushing his teeth became cumbersome and messy.

A week passed, and Joe became more clumsy. He stopped being able to button his shirt. Tying shoelaces became impossible. Trying to use a tablet one morning, his finger swayed from side to side, tapping things to the right or left of what he intended. Lost in an application he had never meant to open and could not figure out how to close, he put the touch pad down on the speckled kitchen counter and gave up. He would never pick up the technology again.

When his wife confronted him about his unsteadiness a few days later, Joe said it was overkill to see a doctor. He could not conceive of having a serious health issue; he had never even had a mild one. Nevertheless, he admitted he could not explain his newfound difficulties. He was not sure why his body trembled.

So Joe agreed to go to the emergency room. He pulled his jacket out of the closet and shimmied the sleeves up to his shoulders. He reached for the zipper, swatting at the bottom of the jacket with imprecise movements. He missed, then missed again. His wife, now even more confident something was wrong, took hold of the zipper and fastened it for him.

At the hospital, Joe's blood and urine studies could not explain the symptoms he was having. The doctor suggested another day of testing and observation. He wanted to admit Joe to the hospital ward.

Soon, Joe lay on a gurney as a technician moved him, head first, into a giant, tubular magnet that banged loudly. On a computer screen in the next room, an image of his brain appeared. The picture showed signs of damaged brain tissue, the first abnormal finding among his many tests. Along the periphery of his brain, a string of neurons that were supposed to appear gray on the images were instead as white as printer paper.

Days later, still in the hospital, Joe had more trouble walking. He needed someone to hold him up by the arms as he tried to lift his feet off the ground. Physical therapy became harder instead of easier. His speech became tangential, traipsing from topic to topic without a clear end point. He mixed up words like *fork* and *spoon* and syllables like *bike* and *bite*. His dry wit, which had entertained nurses just days earlier, gave way to near silence. He stopped asking when he could go home.

Joe's mind continued to flounder as his hospitalization stretched through a second week. When a doctor tested his thinking, he performed like someone with dementia. He could not follow a basic pattern or copy a drawing of a cube. Most alarming for his family, he could not recall a set of words that the doctor had asked him to remember just a few minutes earlier. His recollections were disappearing.

Within days, a technician in gray scrubs rolled a computer and a clump of tangled wires into his hospital room. Using a foul-smelling adhesive, she attached two dozen electrodes to Joe's scalp. His hair stood in all directions, as if in fright, to make room for the small pieces. To each electrode, the technician attached a wire that ran to the computer.

The wires detected electrical activity in Joe's brain. On the computer, a tracing began traveling across the screen. The line dipped down, then swooped upward, rising tall before turning around and careening downward yet again. The shape recurred rhythmically, once or twice per second, like the beat of a metronome. The waves were virtually diagnostic: Joe had a deadly condition called Creutzfeldt-Jakob disease.

Doctors called Joe's wife into a conference room to disclose the news. His family had never heard of Creutzfeldt-Jakob disease before. "What is that?" they asked, adding, "Can you write it down for me?" Through pitying staff and horrifying internet searches, they began to see that Joe was dying.

The following day, a palliative care doctor knocked on the door of Joe's hospital room. "Let's talk outside," his wife said, signaling to the hallway. "I don't know how much he understands." Along with the neurology team, the doctor reviewed Joe's prognosis. With the rapidity of his decline, he was likely to die within weeks. His wife understood that placing a feeding tube would not bring him back to vigor. He would not live long enough to benefit from the medications his nurses woke him up to take. There was no point to more blood draws.

When Joe left the hospital a short time later, he lay on a stretcher that was loaded into an ambulance that took him to a hospice facility. Soon, no longer able to recognize his wife or sit up in bed, Joe died from Creutzfeldt-Jakob disease. The entire ordeal—from gracefully retired to buried—lasted only a month.

~~~~~~~

IN 1921, A GERMAN NEUROLOGIST named Alfons Maria Jakob—a close friend of Alois Alzheimer—published four papers describing a set of patients who suffered an unfortunate constellation of symptoms. Over just a few months, the men and women had developed profound difficulties with movements, language, and behavior. Reaching out for an object, their hands would swing from side to side, completely missing the intended target. Walking became a precarious endeavor as their legs shook underneath them. Simple facts escaped them, so they could no longer tell where they were or what had happened the day prior. Eventually, they lost the ability to write and speak. Despite the impairments, many of the patients behaved with jocularity, as if they were attending a celebration and not witnessing the disintegration of their faculties. Within months of getting sick, most of the patients died.

The syndrome acquired the name Creutzfeldt-Jakob disease after Jakob incorrectly gave credit to a fellow German scientist, Hans Creutzfeldt, who Jakob felt had described a similar case a few years earlier. Creutzfeldt's case turned out to be a different disease—and the pairing ended up being particularly unfortunate, since Jakob was Jewish and Creutzfeldt likely had ties to the Nazi regime—but the name stuck.

By the late 1950s, Vincent Zigas and Carleton Gajdusek understood that the connection between kuru and Creutzfeldt-Jakob disease was not limited to the microscopic. The maladies caused similar symptoms, too. Both conditions inflicted an unsteady gait and shaking arms. People with each disease lost the ability to speak and understand words, then progressed rapidly to death, usually within a year. Perhaps, Zigas and Gajdusek wondered, whatever caused Creutzfeldt-Jakob disease could also be causing kuru.

Then a third disease was added to the mix. In 1959, Zigas and Gajdusek created an exhibit about kuru at the Wellcome Medical Museum in London. Just before the show closed, a veterinary scientist stopped by the display and found that there was something familiar about the microscopic photos of kuru. The pictures looked exactly like the brains of sheep who had died of a disease named scrapie.*

Like kuru, scrapie caused its victims to tilt from side to side. Within months, the sheep would become so unsteady that they could not walk or even stand upright. Then, less than a year after they first became sick, the sheep would die.

The veterinarian wrote a letter to Gajdusek—and to a prominent medical journal—detailing the microscopic similarities between sheep with scrapie and humans with kuru. He

---

* Scrapie was so named because it caused sheep to scrape their wool incessantly against barn walls and fences, leaving behind bare patches of raw skin.

described how blood and urine tests in both diseases were normal, and how scientists had failed to isolate a causative organism in either condition.

Gajdusek, who had just returned to the United States, knew exactly what to do when he read the veterinarian's note. Scientists had already proved that scrapie was infectious, so Gajdusek set out to show that kuru and Creutzfeldt-Jakob disease were transmissible, too. Carefully, he cut out pieces of brain tissue from people who had died of either disease. He used chemicals to turn the samples into liquid. Finally, he injected the material into several chimpanzees that lived in his laboratory. Months later, the animals developed symptoms akin to the clinical syndromes in humans. The experiment illustrated that kuru and Creutzfeldt-Jakob disease—like scrapie—could be "caught."

Gajdusek even found microscopic proof that an infection was at play. Under magnification, brain tissue from the chimpanzees looked just like that of people who had died of kuru or Creutzfeldt-Jakob disease. Patches of neurons had melted away, leaving the moth-eaten look of a slice of sponge. The appearance soon became the trademark microscopic feature of scrapie, kuru, and Creutzfeldt-Jakob disease, garnering the name "spongiform encephalopathies," a term still used today.* For his work, Gajdusek won a Nobel Prize in 1976.

Zigas, in the meantime, retreated from studying kuru and

---

* The word *encephalopathy* is derived from the Greek words meaning "brain suffering."

returned to more general public health work due to reasons he would later refer to cryptically as "circumstances beyond my control."* He moved to Australia, where he published two books about his experience in Papua New Guinea and died in 1983 at age sixty-three. Still today, he is far overshadowed by Gajdusek and rarely receives credit for his endeavors.

———

Once Gajdusek proved that kuru was infectious, it was not long until investigations revealed the epidemiological secret of the disease: the Fore had a tradition of eating the bodies of deceased tribespeople. When a corpse was ready to be eaten—often after it had attracted maggots during a temporary underground burial—a Fore tribeswoman would dismember the figure with a bamboo knife and open the skull. Women eager to honor the deceased would consume a slice of the dead person's brain. Leftover pieces were brought home to feed children. Exactly how kuru started is still under debate, but the way it spread is now well accepted: kuru traveled from woman to woman, and woman to child, in the brains of those who had already died from it.

Another anthropological detail explained why kuru decimated the Fore population while leaving surrounding tribes unaffected. The Fore tribe practiced endocannibalism—eating members of one's own group—but neighboring tribes reserved

---

* The exact reason for Zigas's retreat from the study of kuru remains unclear. By this point, research into the disease had become a biological challenge, so it is possible that his training in public health simply became less relevant.

the practice for enemies who had been captured and killed. This meant that other tribes did not contract and multiply diseases already common to their group. Once cannibalism was outlawed in Papua New Guinea in 1960, rates of kuru declined significantly. No new cases have appeared for decades.

As kuru became extinct, research on spongiform encephalopathies continued to expand. Scientists discovered that whatever was causing the diseases had herculean traits. Researchers could heat infected brain tissue to 680 degrees Fahrenheit for an hour, and still the sample could spread disease. They could destroy all the DNA in the samples, but still the tissues would remain dangerous. A slice of brain from someone with Creutzfeldt-Jakob disease could even be squished on a slide for years, all the while retaining its ability to kill someone who cracked the glass pieces open and exposed their contents.

A decade after Gajdusek accepted a Nobel Prize for his work on kuru—and a decade before he would fall from glory over allegations of child molestation—a young neurologist in San Francisco figured out the cause of spongiform encephalopathies. Dr. Stanley Prusiner was a former Eagle Scout who had grown up in one of the few Jewish families in Des Moines, Iowa. Although he moved to heavily Jewish Philadelphia for college, the feeling of being an outsider trailed him as he became a biochemist, championing theories that mainstream scientists considered outlandish. Prusiner had no qualms about challenging the status quo—a trait that made him the perfect person to study a disease that did not conform to it.

To figure out what caused spongiform encephalopathies,

Prusiner used the process of elimination. He separated samples of infected brain tissue into smaller and smaller fractions. With each step, his team of scientists would check whether the remaining sample was still infectious. The process was akin to sifting flour through a set of sieves with ever-smaller openings. In the end, Prusiner found the purified sample was made of something much tinier than a bacterium, virus, fungus, or parasite. It was made only of protein.

To the myriad traits that proteins were known to exhibit, Prusiner added something astounding: proteins could be infectious. He coined the word *prion* (*PREE-yon*), a combination of *protein* and *infection*, to refer to the molecules that caused spongiform encephalopathies. The discovery soon became his life's work. In 1997, he won a Nobel Prize for describing prions—the second such award to be given for the study of spongiform encephalopathies.

Prusiner's data was convincing, but it raised more questions that puzzled scientists. For prions to cause infection, they had to be able to multiply. Consider the growth of fluffy mold on a petri dish or the thick collections of bacteria that crowd the lungs in pneumonia. In each case, an organism that is initially small in number is able to grow and divide, yielding globs of infectious material. But proteins cannot reproduce the way bacteria and fungi can. Without the ability to replicate, prions should have been diluted into oblivion.

A second mystery was even more perplexing. In autoimmune diseases, doctors can often make the diagnosis by identifying a protein that should not be present. Lauren Kane (Chapter 4) became sick because a tumor tricked her body into creating a protein that attacked her NMDA receptors. Mike Bellows (Chapter 5) became ill because his body produced a protein that blocked his glycine receptors. Most people are free of these diseases because they do not have the abnormal proteins that cause them.

The same cannot be said of spongiform encephalopathies. Every human in the world walks around with prion protein in their brains, but only one in seven thousand deaths—like Joe Holloway's—is brought about by Creutzfeldt-Jakob disease. Prusiner knew there had to be another quality of prion protein that determined whether or not it would cause disease.

He was right. In a series of complicated biochemical experiments, he showed that prion protein adopts a different shape in healthy people than in those with spongiform encephalopathies. Normally, prion protein extends in stiff helices, like Slinkys that have been stretched and then frozen in place. In people with spongiform encephalopathies, the helices unfold and the protein crumples onto itself.

Then, something striking happens: the distorted prion protein forces neighboring normal prion proteins to adopt a toxic shape. This is the key to how kuru, Creutzfeldt-Jakob disease, and scrapie take root. Misfolded prion proteins cannot replicate, but they can do something just as effective for causing infections—they can cause other proteins to mimic them. Like a room full of mousetraps where one contraption snaps and then the others follow, a few misfolded proteins can cause an entire brain of normal prion proteins to adopt the dangerous shape.

Scientists now widely accept that proteins can be infectious. Researchers have identified other spongiform encephalopathies, like mad cow disease, which terrorized meat eaters in the United Kingdom in the late 1980s, and fatal familial insomnia, a rare condition that causes people to suffer excruciating sleeplessness that culminates in an early death. There is even hope that prion diseases may soon be treatable. In 2019, scientists showed that a DNA-like molecule could double the life span of mice infected with prions.* The approach has not yet reached human subjects, but it is one of the most promising developments yet for a disease that strikes at random and remains invariably lethal.

Prusiner has begun to argue that prion proteins also cause more common neurodegenerative conditions like Alzheimer's disease and Parkinson's disease. Most scientists still consider the idea to be on the fringe, but Prusiner—yet again finding himself an outsider—is betting on prions.

~~~~~~~

Joe Holloway's widow still lives in the beautifully decorated home she used to share with him. The living room is immaculate, the kitchen table is spotless, and the dishes are always put away. Everything is in its place, except for Joe.

---

* This work was done by Dr. Sonia Vallabh, a researcher at the Broad Institute in Cambridge, Massachusetts, whose mother died of a rare form of prion disease that is caused by a genetic mutation. When Vallabh found out that she, too, carries the genetic mutation, she and her husband left their careers in law and engineering and retrained in biochemistry. Today, the husband-and-wife pair run one of the most productive prion research programs in the country.

Sometimes, his widow catches herself glancing at the reading nook, imagining him working on a sudoku puzzle, looking up at her. She can picture him, pen in hand, perched on the gray upholstered chair with the legs that look like Möbius strips. Once she had imagined their lives together in the same shape, without beginning or end. Now, her narrative is heavy with longing.

If you call Joe's house when she is out, you will find nothing amiss. The answering machine still plays a cheery recording of his voice from long before he was sick. He sounds confident, warm, and fatherly. It is a nice way to be kept, forever youthful, welcoming guests long after you are able to host them.

Part Three

BRAIN INVADERS AND EVADERS

UNTIL NOW, THESE PAGES HAVE INTRODUCED you to some of the largest molecules in the human body. We saw the wrath of DNA, our most enormous molecule. We witnessed the torture of proteins, which are so gargantuan that scientists in the nineteenth century doubted that a single molecule could be so large.

To tell the whole story of the hijacked brain, though, we need to include a group of molecules far less impressive in size but equally threatening to cognition: small molecules. The term refers to those that are no more than fifty times the size of a water molecule. Researchers invented the name to delineate between molecules tiny enough to pass in and out of cells freely and those big enough not to. The average protein is fifty times bigger than the cutoff. The smallest human chromosome is six hundred thousand times even bigger than that.

Small molecules can inflict damage to the brain in two ways: they can be either missing and needed, or present and harmful. I think of the first group as *evaders*, since we depend on having them around and we suffer when they are missing. Vitamins are a classic example of such molecular evaders; they are so critical to our health that we can develop symptoms when we don't have enough of any one of them. The second group of small molecules, which I think of as *invaders*, includes environmental toxins, illicit drugs, and pharmaceuticals that are not normally part of the human body. These molecular intruders change the mind by breaching the brain and altering the way our neurons fire.

The idea of molecular evaders dates back to the early 1800s, when a French scientist named François Magendie proved that *what* we eat is just as important as *how much* we eat. Magendie was a world-famous neuroanatomist known for dissecting animals while they were still alive, often while they were nailed down to his operating table.

In his foray into nutrition science, Magendie's procedure was somewhat less gruesome, though no less deadly. At the time, scientists suspected that humans needed to consume nitrogen to survive, but there was little data to prove it. So Magendie bought a dog and fed it exclusively sugar, a food thought to be nutritious but known not to contain nitrogen. The canine subject lapped up the diet for two weeks. Then, it began to suffer. It lost much of its body weight in a matter of days. Ridges of ribs appeared under its fur. An ulcer developed on the surface of one of its eyes. Just a month after starting the monotonous diet, the dog died. Magendie went on to perform the same experiment with other foods, like olive oil and butter, that also did not contain nitrogen. Each time, the dogs could not survive.

Magendie and his colleagues were quick to assume that the dogs had died because they did not eat enough protein, the nitrogen-rich molecule discovered by Antoine Fourcroy in the midst of the French Revolution. But the fault in that interpretation is easy to spot today: each of the foods Magendie had chosen was not only deficient in protein but also lacked other vitamins and minerals we now know to be essential to survival. The dogs had died not from protein deficiency alone but rather from a lack of several critical dietary elements.

It took decades for the scientific majority to believe that proteins were not the only molecules needed for a healthy diet. By then, Magendie was dead, and his experiments were forever inscribed in legislation against animal cruelty.

Scientists did eventually realize that there had to be more than one essential nutrient. Scurvy, known to be caused by an inadequate diet, was cured by citrus juices that had no protein at all. An intrepid physiologist and chemist showed that they could climb a mountain in Switzerland after eating a low-protein meal that could not alone have supplied enough energy for their expedition.

It was Casimir Funk who finally redefined the field in the early twentieth century. Funk was a reserved scientist whose research spanned six countries as he struggled to escape rising European anti-Semitism. Born in Warsaw, Poland, he was rejected from an illustrious government school because of his religion, but went on to graduate at the top of his class from a private school. By age twenty, he had defended a doctoral thesis in chemistry.

In his most famous paper, Funk wrote about conditions that arose when a person lacked a particular dietary nutrient. He suspected there were several such diseases, each caused by a deficiency in a particular, nitrogen-containing molecule. In homage to their shared atomic makeup, he named the nutrients that were essential for human health *vitamines*, a blend of *vital* and *amine*, the latter term referring to an arrangement of nitrogen and hydrogen atoms that he thought was common to all such molecules.*

Funk argued that the lack of each vitamin could cause a particular constellation of symptoms. He proposed that several diseases whose cause remained elusive were in fact precipitated by a deficiency of an as-yet-unidentified vitamin, a theory that other scientists would later prove to be correct.

Thanks in part to Funk's enthusiasm, hordes of researchers flocked to nutrition science in the first few decades of the twentieth century. Together, they discovered thirteen vitamins in just thirty-five years—a remarkable pace, particularly because progress stalled twice in the middle due to world wars. Funk himself nearly became a Holocaust victim. Just months before Nazis occupied France, he fled Paris with his family, leaving his entire collection of chemicals, labware, and books behind in Europe. He spent the rest of his career in the United States and died in New York City in 1967.

We now know that, just as Funk suggested, each of the thirteen essential vitamins causes a particular syndrome in its absence. Vitamin B12 deficiency—often caused by weight-loss surgery or vegetarianism—manifests with a smoldering confusion, limp limbs, and difficulty sensing where the arms and legs are in space. Vitamin B6 deficiency, usually a side effect of

---

* This turned out to be wrong. Many of the vitamins subsequently identified did not contain nitrogen at all. The name *vitamine* was thus shortened to *vitamin* in order to remove the reference to amine groups.

medications, can cause seizures, dementia, and confusion. Lack of vitamin B1 can induce the mind to create false memories. Vitamins, so vaguely understood until Funk's era, are now one of the most widely studied topics in nutrition science.

～～～～

THE IDEA OF VITAMINS and other molecular evaders dates back only to the past few centuries, but humans have long believed in the existence of molecular invaders. We have known for years that swallowing, smoking, or inhaling so-called medicines could affect our health. Use of medicinal plants began in the Paleolithic period sixty thousand years ago, if not earlier. Documents written in 1550 BCE raise the idea of taking pills to induce healing. Even the first synthetic medication was produced well over a century ago.*

Today, half of all people in the United States use at least one prescribed medication. A full 12 percent fill more than five prescriptions per month. Among these drugs, 90 percent are made of small molecules. While biologics—medications constructed from proteins and DNA-like molecules—have gained increasing attention in the past decade, they remain a minor and more unruly portion of the drug market. Compared to protein and DNA, small molecules are more heat stable, simpler to purify, and easier to package into pills—all characteristics that make them more convenient to manufacture and distribute to patients.

Many of these small molecule medicines—these invaders—

---

* The medication, called chloral hydrate, is a sedative that is still used today.

can have drastic consequences in the brain. A small molecule used to treat the tremors of Parkinson's disease can cause compulsive gambling, stripping victims of their assets just weeks after starting the prescription. A small molecule meant to treat seizures can cause word-finding difficulties, leaving language perpetually at the tip of the tongue. Even the active ingredient in many over-the-counter sleep agents can hinder thinking skills and may increase the risk of developing dementia.

Since most people do not advertise their medication lists, we get to know each other without thinking of how the small molecules we ingest manifest in our interactions. We do not often consider what arguments, love stories, and tragedies came into existence courtesy of the pills we swallow every day— the molecular invaders we chose to bring into our brains. We become one with our pharmacotherapy. The small molecules we consume become part of our identities.

Chapter Seven

LIKE LUCIFER

EVEN FOR A LAWYER AS SEASONED AS ABRAHAM Lincoln, *The State of Illinois v. "Peachy" Quinn Harrison* was a daunting case. Brawling in a local drugstore in 1859, Harrison had thrust a four-inch blade into his longtime rival, Greek Crafton. The white-handled knife tore into Crafton's skin just below his ribs, then dragged across his belly to his groin. He died three days later. Harrison went into hiding, stowing away under the floorboards at a local college, until police took him into custody.

Called to defend Harrison, Abraham Lincoln—in the last case he would take on before rising to the presidency—set to work exploring details of the story that might help his client: Harrison had been sitting at the counter, reading a newspaper, when Crafton entered the store in a rage. Crafton grabbed Harrison and tried to drag him to the back of the store for a beating, but Harrison twisted and threw Crafton off balance. Both men stumbled into a mound of boxes. It was at that moment, without an obvious exit, that Harrison pulled out a knife and killed his attacker.

Lincoln's case depended on the testimony of a local reverend who said he witnessed Crafton make a deathbed confession taking responsibility for the conflict. But when the reverend took the stand in the early fall of 1859, the prosecution objected. "Dying declarations are not admissible evidence," the lawyers argued. The judge agreed.

Lincoln's voice erupted from the defense table. "Your honor, we need to see this through, every last bit of it!" His words were such a surprise, such an upset of typical courtroom protocol, that even the stenographer looked up, openmouthed. Calm and cool-headed Abraham Lincoln had been lit on fire. According to one courtroom witness, Lincoln rose from his chair "like a lion suddenly aroused from his lair." He barreled toward the bench with such fervor that those in attendance wondered if he was going to climb right over the barrier and onto the judge. Lincoln became indignant, berating the court and the man who presided over it. "It was a display of anger, the likes of which I never saw exhibited," said one observer.

"The deceased has a right to be heard," Lincoln barked, flailing his long arms.

"You finished?" the judge asked, hoping to retake control of the room.

"I am, your honor," Lincoln replied, regaining his composure and returning to his seat. "Thank you."*

It is difficult to imagine "Honest Abe," the unflappable genius of a president, with the disposition of a baited bull. A colleague from the White House would later say he had "never

---

* Lincoln went on to win the case, his final trial before assuming the presidency.

heard him speak a word of complaint." Others who had worked with Lincoln during his presidency found he "was always of even temperament, never showing anger." This is why most people imagine him with the same rock-like stoicism as the statue that memorializes him in Washington, DC. For most of his time in the White House, that was his modus operandi.

But some of those familiar with Lincoln before his presidency knew a different character. His first law partner wrote that "as he was grand in his good nature, so he was grand in his rage." Another of his partners remembered that "it did not take much to make him whip a man." Lincoln at times appeared "so angry that he looked like Lucifer in an uncontrollable rage."

In the fall of 1858, Lincoln ran for senator of Illinois against the incumbent Stephen Douglas. The tension between the candidates was so intense that when they agreed to engage in live debates, people came as much for the biting banter as for the political discussion. Lincoln remained stately and calm throughout the first three showdowns. He was polite, pensive, and senatorial.

And then he was not.

During the fourth debate, the two men sparred over whether Lincoln had supported the troops during the recent Mexican-American War. Looking for someone who could vouch that he had backed the soldiers, Lincoln searched the podium and recognized a Douglas supporter who had served in Congress with him. Pointing his finger at his old colleague, Lincoln commanded the man to testify on his behalf.

Before words could be uttered, Lincoln wrapped his hands around the man's neck and lifted him upward. According to one account, he hauled the man forward "as if he had been a kitten."

The man's feet dragged along the platform floor. His teeth chattered. Lincoln's long fingers suddenly appeared as a noose; the white fabric of the man's coat collar folded under his grip. For many in the auditorium, the possibility arose that they might witness a murder of passion and not simply a political debate.

Then one of Lincoln's own bodyguards stepped forward and pried his charge's hands from the victim's neck. The man slumped down, heaving. The crowd moved from astonishment to laughter, as if the last few minutes had been a planned performance and not a display of unregulated emotion. Lincoln went forward with his argument, once again showing the calm demeanor he had displayed in previous debates.

In explaining Lincoln's temper, his law partner once said that "when a great big man of mind and body gets mad, he is mad all over, terrible, furious." What the partner did not consider— what he could not have conceived of, given the science of the time—was that the rage could have been biologically instigated by a molecular invader. In moments of uncontrolled wrath, argues world-renowned doctor and medical historian Norbert Hirschhorn, Lincoln's mind may have been under the spell of a medication called blue mass.

The idea first occurred to Hirschhorn while he was reading Gore Vidal's historical novel *Lincoln*. In the book, Vidal describes an encounter between a tavern owner and a pharmacy clerk near the White House. Inebriated, the clerk boasts about his celebrity clients. "His bowels don't ever move properly," he says of Lincoln, "so we fill him up with blue mass."

Most physicians in the 1990s did not know what blue mass was. It had long been removed from hospital formularies and pharmacy shelves. It could not be purchased from vitamin stores. But Hirschhorn, by then on the verge of retirement, recalled the name from his time in medical school half a century earlier. The main ingredient in blue mass, he remembered, was mercury.

～～～～～

MERCURY IS A MESMERIZING element. It flows like water at room temperature and atmospheric pressure, a trick no other metal in the periodic table can perform. If you separate a bit of mercury from the rest, it beads up in tantalizing mounds. With a nudge, it rolls along a flat surface like a drop of water. Gold and platinum are beautiful in static form, but mercury's mystique lies in its dynamism, in the way it makes people question the laws of physics.

Mercury was used as medicine for millennia before the sixteenth presidency of the United States, and for many years after Lincoln was assassinated. It was only in the twentieth century, with advances in epidemiology and biology, that humans discovered mercury's dirty secret: it is as dangerous as it is beautiful.

In 241 BCE, the first emperor of the Qin dynasty took mercury as an antidote to mortality. The silvery substance likely had the opposite effect; historians believe the emperor died of mercury toxicity at age thirty-nine. In preparation for his demise, thousands of people spent more than a year building an elaborate resting site laced with rivers of mercury. The outer portion of the area has become known for its famous terra-cotta warriors, but the mausoleum they protected remains unopened today. The structure is so encased in mercury that nobody has dared to excavate it.

In the early 1800s, Lewis and Clark carried a mercury concoction called "thunderclappers" in their trek across the United States. The pills were prescribed to expedition members for everything from syphilis to yellow fever. Doctors at the time believed mercury purged the body of toxins by inducing voluminous diarrhea—a side effect that turned out to have little therapeutic benefit but was fortunate for modern-day archaeologists; mercury deposits, still present two hundred years later, mark the sites of the group's latrines.

By the early twentieth century, calomel—another mercury-containing medication—became a common ingredient in teething powders. Parents rubbed the product into their children's gums when the white sheen of a new tooth poked through. As a result, children's fingers and toes became swollen and painful. Their skin sloughed off. Doctors called the condition "pink's disease," after the irritated color of raw flesh. When they discovered that the condition was caused by mercury, companies took calomel out of baby products. The epidemic disappeared, and parents were left with the horrifying realization that they had caused their children's pain.

Today, most people associate mercury with old thermometers more than with old medications. A block of gold will not change noticeably in volume if the ambient temperature is raised by two degrees, but a column of mercury will expand by an amount that can be seen with the naked eye. If the temperature is lowered again, the liquid visibly shrinks, making it the ideal substance for seeing temperature change—unless the thermometer cracks.

When mercury is left to its own devices, it can transition from a liquid to a vapor at room temperature. In its aerosolized form, it is odorless and colorless, a stealthy toxin that makes its way past the body's defenses by floating into the nostrils and the mouth. From the respiratory tract, it infiltrates nearly every organ and tissue, depositing in the heart, liver, pancreas, lungs, and thyroid. It invades the salivary glands, inducing such a powerful response that a Swiss doctor in the sixteenth century claimed that a therapeutic dose of mercury would produce three pints of saliva per day.

The brain is mercury's most vulnerable target. There, it opens us up to the ravages of dangerous forms of oxygen. It changes the balance of calcium inside and outside of neurons, so the cells fire when they are supposed to be silent. Some forms of mercury even destroy the structural proteins that give neurons their long shape, rendering the cells unable to communicate with each other. Most strikingly, mercury convinces neurons to carry out a massive act of cellular suicide. Mercury—once thought to be a panacea for everything from syphilis to depression—actually tricks the nervous system into sabotaging itself.

The clinical effects of mercury toxicity are just as dramatic. Children exposed to mercury in the mid-1900s erupted with wild behavioral changes, alternating between complete apathy and extreme anger. Many became insomniacs, staying awake for days at a time as their moods flared. Depression, and even hallucinations, sometimes took hold.

In 1988, a botched pipe replacement at a factory in Tennessee unleashed a massive mercury leak that made headlines.

People exposed to the heavy metal suffered hot tempers and heavy fatigue. They stopped socializing and retreated into isolation. Many had outbursts of violence similar to the ones Lincoln experienced more than a century earlier.

Today, acute mercury poisoning is rare in developed countries, largely because of workplace regulations. In 2019, less than 1 percent of calls to poison-control centers in the United States pertained to questions about mercury. Most cases of modern-day toxicity arise from overzealous consumption of tuna, mackerel, and other marine animals that contain high levels of methylmercury, a molecule made of mercury, carbon, and hydrogen atoms. Because many of these fish are expensive, mercury toxicity has in some ways become a disease of celebrity.*

~~~~~

TO FIGURE OUT WHETHER Abraham Lincoln could have absorbed enough mercury to experience toxicity, Norbert Hirschhorn recruited a colleague with expertise in pharmaceutics. The team turned to an 1879 compounding textbook, a tome of more than 1,600 pages that contains recipes and indications for now-debunked medicines like arsenic (for leprosy, psoriasis, and malaria) and gold (for tuberculosis). Halfway through, the text lists *Massa Hydgrargyri*, or "mass of mercury," along with a

---

* The British singer Robbie Williams, *Hidden Figures* actress Janelle Monáe, and *Entourage* actor Jeremy Piven have all spoken about suffering symptoms of mercury toxicity after adopting a pescatarian diet.

recipe for blue mass. "I think we should remake it," Hirschhorn said.

According to the recipe, blue mass was one-third mercury by weight. To this, a pharmacist would add licorice root, glycerol, rose water, honey, and the petals of a hibiscus plant. The mixture was pummeled between a mortar and pestle until the metallic globules of elemental mercury disappeared. The paste was then rolled into a strip and sliced into pieces that could be shaped into pills.

Hirschhorn's collaborator became like a chef in search of rare ingredients. She purchased mercury from a biochemical supply company. She bought honey from a grocery store. She procured hibiscus flowers from a local florist.

In preparation for the experiment, she donned a surgical gown, gloves, and a mask to fend off mercury vapors that could emerge as she pounded the ingredients with a mortar and pestle. She mixed everything under a fan that pulled airborne particles upward, instead of releasing them into the laboratory. Only when she had created a paste—once the mercury was integrated with the rest of the ingredients—did she roll the mixture into a thin log and cut it into the shape of tablets.

The pills were ready, but determining what would happen when a person swallowed them remained complicated. A real-life experiment was unsafe and unethical, so Hirschhorn devised an alternative: the pills would be crushed in a sealed bottle with acidic solution to simulate the stomach environment. Then a machine capable of quantifying heavy metals would measure the mercury content in the bottle's airspace.

The results of the experiment suggested that the vapor from

a single blue mass pill contained more than thirty times the known safe limit for mercury exposure. Since most doctors in Lincoln's era recommended taking the pills two or three times per day, the actual exposure from vaporized mercury could have been even higher.

In 2001, Hirschhorn published a paper titled "Abraham Lincoln's Little Blue Pills." The article proposed that Lincoln had stopped using blue mass in the first few months of his presidency, because he realized the pills "made him cross." If the timeline was correct, Hirschhorn argued, it meant that Lincoln was aware that a medication threatened his ability to run the country—and that he had the wherewithal to stop using it before it was too late. "What this may mean to an evaluation of Lincoln's achievements is mind-boggling," a renowned Lincoln scholar said of the theory.

Hirschhorn's work has since become a controversial piece of literature on Lincoln's health. Today, a diagnosis of mercury poisoning is based on proof of elevated mercury levels in bodily tissues—something essentially impossible to illustrate in Lincoln's case. We have no blood or urine samples to test. Hair strands can carry traces of mercury for up to a year, but if Lincoln stopped using blue mass three years before his death, locks cut after his assassination would be useless.

Numerous sources offer secondhand accounts that Lincoln used blue mass, but we have no proof that he ever filled prescriptions for the medication. Ledgers from a Springfield, Illinois, pharmacy showed that the Lincolns purchased 245 items between 1849 and 1861, but none of the products is identified as blue mass.

Hirschhorn believes there was reason for Lincoln to cover up his use of the pills; he suspects the evidence is hidden not because of the erosion of time but because of a deliberate effort. Some sources agree with Gore Vidal's story that Lincoln took blue mass for constipation, but most suggest he took the medication for depression. Lincoln was a lifetime victim of sadness, but his campaign advertisements depicted him as a hearty log-splitter, a go-getter with boundless energy. Publicizing his need for psychiatric treatment could have hurt his bid for the presidency.

Perhaps in support of Hirschhorn's suspicion, an 1861 letter to Lincoln from another pharmacist in Springfield reads, "I now hasten to send you the Pills as requested." Experts do not know the identity of the pills in question or the reason Lincoln purchased them from a separate dispensary from the one he usually used, but Hirschhorn thinks the answer has to do with blue mass.

A final piece of evidence remains purposefully hidden. As part of his research, Hirschhorn found an article in an Illinois newspaper about Brownback Drug Company, a family-owned pharmacy twenty miles outside Springfield that reportedly filled a prescription for Abraham Lincoln. Fortunately for Hirschhorn, the pharmacy was still in existence in the 1990s.

Wondering why Lincoln would purchase prescriptions from a pharmacy just outside his hometown, Hirschhorn wrote a letter to the modern-day pharmacy owner asking if the prescription was for blue mass. Citing patient confidentiality, the owner wrote back a cryptic note that Hirschhorn interprets as veiled confirmation: "The fact that Lincoln chose to seek medical and pharmaceutical care outside of his immediate community certainly acts to heighten our sense of responsibility in the matter," he wrote to Hirschhorn. "While I sincerely regret that I am unable to further assist in your research, I do advise that the various mercuric compounds were common ingredients in many mid-nineteenth century prescriptions."

Still on the case over a decade later, Hirschhorn reached out to the man again. The pharmacy had since closed, and its former owner was working in college administration. Hirschhorn invited him to an upcoming lecture on Abraham Lincoln and included a reference to the published paper on Lincoln and blue mass. He hoped the man might finally reveal the identity of the prescription.

There was no such turnaround. The speaking engagement passed without the man's attendance. "The most fun about my various researches," Hirschhorn later reflected, "was in chasing down different rabbit holes and every now and then finding a rabbit." As for the unnamed prescription, he is still on the case.

Chapter Eight

An Honest Liar

By 2018, Lisa Park's memory had acquired a curious flaw. Instead of being a place to store accounts of recent events, her mind became like fertile soil that grew tales of incidents that never occurred. It was a tool for creation, rather than a mechanism for recollection. Lisa herself was unaware of the change and had no idea her memories were false. She had unknowingly become an honest liar.

As a child in the 1960s, Lisa had fixed cars even before she was old enough to drive them. The daughter of a middle-class mechanic, she was raised around cylinders and pistons. By the time she was fifteen, she had visited racetracks in every state in the contiguous United States. She wore a leather jacket and a short haircut as she worked on engines with her father until late in the evenings. She regularly slept in tents and trailers, her hands covered in oil and roughened from metalwork.

When Lisa turned sixteen, her mother commanded, "You have to become a girl." She said car racing and auto culture were unbecoming of a young woman. The leather jacket had to go. Her short hair was to be grown out. Her hands were cleaned, moisturized, and manicured.

Resistant at first, Lisa eventually adopted a feminine edge with the same fierceness she had brought to cars. She became a fashion devotee. Silk blouses took the place of ripped T-shirts. Jeans were replaced by dress slacks and stilettos. A faux-mink coat entered her wardrobe. Wherever she went, whether it was to the grocery store or the orchestra, she was dressed spectacularly.

So it is no surprise that Lisa was wearing four-inch pumps when she met Johnny in one of the aisles at their local grocery store in 2002. Lisa was forty at the time, divorced with no children, intending to grow old surrounded by her eight Dalmatians. Johnny was a father of four, a gregarious widower who thought he was too homely to find another woman to marry him. After dating for a time, the pair married in an impromptu ceremony performed in Jamaica.

For the first fifteen years of their relationship, Lisa took care of Johnny, his children, and their collection of dogs. She practiced algebra with Johnny's youngest son, who had been failing classes ever since his mother died. She convinced Johnny's daughter to stop skipping school. She stocked the refrigerator with healthy food, and she encouraged Johnny to exercise. Speeding around town in a red roadster, she was forever a woman on a mission.

Then, in 2015, Lisa's feet stopped working as they used to. Instead of lifting up from the ground as she walked, her toes hung low and tended to trip her. Walking into a party store to buy Halloween decorations, her foot caught a ripple on the gray entrance rug. She fell, fracturing several ribs.

Within a month, tingling sensations crept from her fingertips toward her shoulders. The prickling feeling of pins and needles kept her awake at night and distracted her during the day. She tried taking hot showers and cold showers, applying menthol cream and taking pain relievers. The harsh sensations spread, invading her lips and then her tongue.

Whatever was happening, she soon realized she could not fix it with her usual toolbox of wrenches and stoicism. Her ankles weakened, followed by her knees. She resorted to counter-surfing to move around her house, asking Johnny to pull chairs and couches into the middle of rooms so she could transfer from one surface to another without having to stand by herself.

Finally, Lisa traded her faux-mink coat and designer slacks for a stiff hospital gown that tied loosely in the back and tended to leave her exposed. Over the next twenty-eight days, doctors ordered brain scans, spinal cord pictures, a lumbar puncture, and a nerve study to get to the root of her ills.

Eventually, the physicians concluded that Lisa's immune system was the problem. Listening intently to their jargon, Lisa gleaned that her body had attacked the shiny sheaths that surrounded her neurons. The covering was critical for an electrical signal to travel from one end of a neuron to the other. Without it, she understood, her nerves could not communicate with each other.

With Johnny by the bedside, Lisa watched optimistically as clear bags of medications emptied their contents into a long tube that led into a blood vessel in her arm. She imagined driving around town in her convertible again, cured by treatments that would reverse the devastation of the prior months.* Thinking her strength would recover, she returned home.

But the medications did not work. A divot began to develop on the large, cushioned chair in the living room where Lisa planted her body from the time she got out of bed in the morning until she crawled back to her room at night. She began using a wheelchair to get around her house. Johnny moved the furniture again, this time hauling couches and side tables to the perimeter of rooms so there was space for the wheelchair to make turns.

By 2018, Lisa was becoming more and more confused. Most days, she forgot she had retired. She would wake up early and beg Johnny to bring her a blouse from the closet, worrying she would be late for work. When she looked at photos, pieces of images crept into her reality. A picture of her new grandson was elaborated into a birth story that had never occurred. An image of a celebrity could turn into a full-fledged story of a brush with fame. Instead of being a place to store experiences and factual information, Lisa's memory became a tool for confabulation. Fictions inserted themselves into her recollections, creating a truth that never was.

Lisa's doctor explained that an autoimmune disease was now increasingly unlikely. It would not have caused her to confabulate and would have improved with the medications she had taken. Distraught by both the loss of an old diagnosis and the lack of a new one, Johnny mentioned a secret he had promised

---

* These were the same medications that Lauren Kane (Chapter 4) and Mike Bellows (Chapter 5) received for ailments caused by toxic antibodies.

Lisa he would not share, a detail he had not realized might be relevant: Lisa had an addiction to alcohol. Empty wine bottles had accumulated around the house over the prior years. Two glasses would turn to five or eight, so he might see his wife sober only a few times a week. She had been eating less as she drank more, shedding pounds until the designer clothing she had once loved began to sag formlessly around her thinning frame.

With the new information in hand, Lisa's neurologist told him to bring her to the hospital immediately. The doctor suspected a vitamin deficiency, a condition first described 150 years earlier by a cerebral Russian psychiatrist named Sergei Korsakoff.

~~~~~

DR. SERGEI SERGEIEVICH KORSAKOFF was born in 1854 in a Russian mining town surrounded by massive quartz and dolomite quarries that fed local glass factories. Korsakoff's father managed one such glass plant, but young Korsakoff had little interest in joining the family business. He was more keen on creating theories than fashioning objects, and he quickly found that his brain was a better tool than his hands. As a teenager, he went to Moscow to attend medical school and never moved back.

Adorned with a doctoral degree, Korsakoff joined the clinic for nervous diseases at Preobrazhenskii Hospital in the northeastern part of Moscow. A burly and serious-looking man, he advocated against forced sterilization of the insane at a time when the practice was widespread throughout much of the world. He threw out straitjackets and took off restraints so that patients could roam freely on the wards. While much of the world saw insanity as evidence of moral failing, Korsakoff viewed mental illness as similar to any other, less stigmatized, disease.

Korsakoff first met Arkady Volkov in June of 1889 in the halls of Preobrazhenskii Hospital. Arkady was a thirty-seven-year-old author with a stereotypical writer's burden of angst and defects. He had long nursed a habit of drinking brandy as an antidote to wordlessness, a practice that would one day make him the subject of one of Russia's most famous medical papers.

According to Arkady's friends, who had brought him to the hospital hoping he would recover, he had begun to need reminders of what to do each day: *Wear this*, or *eat that*. He repeated himself in conversation without realizing he had told the same tale to the same person a short time earlier. It was as if every moment was dropped into time without context. His daily routine had become a novel experience with each sunrise. When confronted with the conflict between reality and his memory, Arkady would become resistant and aggressive. He would deny that his mind had faltered.

Arkady's arms and legs had weakened, too. He acquired the gait of a much older man, walking unsteadily past the pillared hospital buildings. Sharp pains ran up and down his limbs, becoming so excruciating that he would cry out in distress from the hospital ward. By the time a nurse came to him from across the room, Arkady was often confused by the encounter; he could no longer remember calling for help in the first place.

Arkady's writing career suffered greatly. Looking at letters he had received from editors a month earlier, he could not recall reading the documents before. When he reviewed a half-finished essay that he had worked on prior to his illness, he could no longer remember the intended climax or resolution. All that he had not put on paper had disappeared from his memory. "The recent past left no traces on the patient," Korsakoff wrote, describing the profundity of his patient's amnesia.

Arkady could still solve tasks of logic. He could deliver an impassioned diatribe on a controversial topic. He could deal playing cards. He could play checkers with foresight, sliding each piece from space to space while calculating the ways his opponent could respond. As soon as staff removed the board, he would have no recollection of the match.

"The patient remembered nothing of what he had done but told of things that had never occurred," Korsakoff later wrote, describing the most striking feature of his patient's disease. Arkady believed he had written a short story when in fact he had only ruminated on it before he became sick. He would describe the previous day's visit to a far-off place, when in actuality he had been too weak to get out of bed. When two coins—among his few possessions—were lost over the course of his hospitalization, he created a complicated tale of theft in which his brother-in-law had come to him gripping the gold pieces and saying, "Arkady, look here. You'll never lay eyes on these coins again." His memory had been impregnated with shards of truth that grew into full-blown recollections. He had become an honest liar.

Then Arkady's arms and legs stopped moving entirely. It looked as if he was stuck in an imaginary straitjacket, restrained to his bed by gravity. His breathing slowed. His chest stopped rising and falling altogether, and he died, just a few months after meeting Korsakoff.

With the benefit of an autopsy, Korsakoff published an account of Arkady's case. He described how nerves that stretched between Arkady's spine and limbs were destroyed, leaving his brain functionally severed from his arms and legs. This, Korsakoff wrote, was why Arkady had become so weak that he could not lift himself out of bed. The condition, which other doctors had identified before, was called multiple neuritis, for the way it affected so many neurons throughout the body. The cause was unknown.

The aspect of multiple neuritis that had not been explored in depth before—Arkady's profound confusion—became Korsakoff's signature work. He described the way Arkady's brain created false memories, the way he correctly recalled things that had happened before he was sick but had inaccurate memories of subsequent times. Korsakoff's name was soon attached to the strange cognitive profile that sometimes accompanies multiple neuritis. The constellation of symptoms was written into medical history as "Korsakoff's syndrome."

Korsakoff never figured out the cause of the condition that bears his name. He suspected it came from a toxin that floated in alcohol, or a poison the body could make in response to imbibing, but he did not undertake the experiments to prove his theory. Instead, he spent the next ten years of his life advocating for the mentally ill. He suffered two heart attacks at age forty-four, then died of heart failure two years later. In the courtyard of the Moscow clinic where he carried out his life's work, a bust carved in red granite still stands today. "To S. S. Korsakov—," a plaque under the sculpture reads, "Scientist, psychiatrist, thinker, humanist."

~~~~~

AROUND THE TIME KORSAKOFF published his paper on Arkady, a researcher named Christiaan Eijkman took an interest in multiple neuritis. Eijkman was a thirty-year-old scientist who had sailed through medical school with honors and had trained under some of the best biologists of the time.

After joining the Dutch military in the late 1800s, Eijkman was sent to the island of Java to solve an outbreak of multiple neuritis among Dutch troops who were stationed in the area. The disease had spread through entire battalions with shocking speed. Soldiers who arrived for training in good health began to limp after just a few weeks in their units. In one army hospital, eighteen men died of the disease in a single day. Still other outbreaks of the condition occurred in nonmilitary populations, in prisons and on ships, bringing the toll of multiple neuritis to the thousands among Dutch people alone.

Eijkman, like Korsakoff, initially believed a toxin or an infection caused multiple neuritis. But none of the toxins he examined could recapitulate the condition's symptoms. When he looked under a microscope for an organism to blame for the supposed infection, he found nothing.

As it happened, Eijkman kept a flock of chickens to use as research subjects in his laboratory. While his scientific exploits were failing, the chickens—which had not yet been used in the experiments—began to die. The birds did not expire in the usual manner that chickens do, overcome with bacteria or killed off by other chickens. Rather, they died with the same symptoms as the Dutch troops. The chickens became wobbly on their perches and struggled to stay upright. When they could no longer stand on their legs, they would keel over and lie on their wings, which had become so weak that they could not flap anymore. Their respiratory muscles would slow to an eerie rhythm on the cusp between life and death. Then their breathing would stop altogether.

Eijkman wondered whether the chickens could be used as an animal model for multiple neuritis. To see if the condition was infectious, he took samples of body fluid from the dying birds and injected them into healthy birds. Not long after, the normal chickens became ill—suggesting that multiple neuritis was indeed transmissible. But another part of the experiment did not quite fit: control birds, which had not received the injections and had been moved to a separate area of the laboratory, came down with the affliction anyway. No amount of isolation could save them. The finding perplexed Eijkman. It cast doubt on the possibility that multiple neuritis was transmissible, but also failed to suggest an alternative mechanism.

Several weeks after the disease began, the sickness cleared up without any intervention from Eijkman. Chickens that had not died recovered. No new cases occurred among the birds for weeks. Before Eijkman could solve the disease, something had cured it for him.

Eijkman began to suspect that the answer was in the animals' food. He spoke with the laboratory keeper to see what the chickens had been eating. "In June, to save money, we started feeding the chickens leftover white rice from the military kitchen," the man explained to Eijkman, who took note that the disease had appeared just three weeks after the switch. "But then in November, the chef was replaced by a stubborn man who refused to give military rice to civilian chickens." With a tiny budget and an entire coop to feed, the manager returned to feeding the flock brown rice, which was cheaper than the white rice served to soldiers. Within days of switching back to the brown rice diet, the birds had regained their health.

With impressive proficiency for a government endeavor, Eijkman deployed his findings to more than one hundred Dutch prisons with some three hundred thousand inmates. He divided the facilities into two groups and instituted a white rice diet in one and a brown rice diet in the other. Within weeks, cases of multiple neuritis disappeared from places where brown rice was served. Brown rice, long considered the less palatable, impoverished version of white rice, turned out to be a miracle cure for multiple neuritis.

A few years later, in the same institution where Eijkman's chickens had suddenly become ill, a pair of Dutch researchers figured out what brown rice had that white rice lacked. The scientists bought four wooden drums, each so enormous that it reached to their waists. From India, they purchased the discarded pieces—the "dedek"—from factories that polished brown rice into white rice. The dedek, they assumed, contained the nutrient that prevented multiple neuritis.

In a series of complicated biochemical steps, the researchers purified a white, crystalline substance from the barrels of dedek. From two hundred kilograms of the rice remains, the scientists produced just 1.4 grams of the substance they hoped would cure multiple neuritis. Excitedly, they fed the white crystals to sickly birds and waited to see the results—which turned out to be spectacular.

Birds suffering from the avian equivalent of multiple neuritis recovered almost immediately after eating just a minuscule amount of the white substance. The scientists named the mystery factor "aneurin" after its anti–multiple neuritis activity. When researchers later discovered that the crystals contained sulfur, they changed the name to thiamine, a combination of the Greek word for sulfur (*thio-*) and the term *vitamine* that had been coined decades earlier.

Thanks to research that stretched from Moscow to the Netherlands, thiamine became the first vitamin ever purified from food. Its discovery paved the way for researchers to identify twelve other small molecules that together make up the breadth of essential vitamins recognized today.

～～～～

THIAMINE IS A SMALL molecule with big jobs throughout the nervous system. It allows us to synthesize signaling molecules that enable our neurons to communicate with one another. One such molecule, acetylcholine, is famous for being deficient in people with Alzheimer's disease. It is no surprise, in light of this, that people with Korsakoff's syndrome show some of the same memory symptoms as those with Alzheimer's disease.

Thiamine also protects us from dangerous forms of oxygen that would otherwise decimate our brain cells. These so-called "radical oxygen species" have a tendency to destroy our DNA and cripple our proteins, leaving catastrophic cognitive damage in their wake. With thiamine around, we become chemically capable of dismantling the threat. We can defend our faculties against destruction.

When our neurons extract energy from sugar—a critical act, since the brain is the most metabolically active organ in the body—thiamine allows the process to run smoothly. Without thiamine, we struggle to turn bread, pasta, and other carbohydrates into molecules that can power our thoughts.

We now know that drinking heavily can decrease thiamine absorption in our gut and even impair our ability to use the thiamine that does get absorbed. This is what caused Arkady to develop symptoms in the first place. But exactly why thiamine deficiency produces such an astonishing malady of memory—and why the cognitive effects of thiamine deficiency manifest in some people but not others—is still unclear today. The connection from biochemistry to confabulation is still missing a link.

IF YOU OPEN LISA PARK'S electronic medical record and turn to the tab that shows her laboratory studies, you will find the result of a test her doctor ordered in 2018, after learning about her penchant for alcohol. Lisa's thiamine level is written in red letters to signal that it is dangerously low: thirty-four nanomoles per liter, barely half the lower limit of normal. It was thiamine deficiency—and not an autoimmune process—that opened her memory for external input. Antibodies had not attacked her brain. Rather, the wine glasses she refilled had turned her into the victim of a molecular evader. Her false memories arose from the same molecular problem that had afflicted Sergei Korsakoff's patient more than a century earlier.

When doctors saw Lisa's thiamine level, they flooded her with the vitamin. Three times per day, an intravenous catheter infused concentrated thiamine directly into her bloodstream. Within days, her thiamine level had increased to normal. Some of her nerves were too damaged to respond to the influx, but others absorbed the vitamin and began to regrow. She could once again hold a coffee cup and slide her legs to the side of the bed. She could sometimes state the year and the month correctly. She was still far from the energetic whirlwind who raced around town in her red convertible, but there was hope she would find her way back.

Since the average nerve grows only one millimeter per day, Lisa's recovery extended over years. It took three months for her to swallow well enough to maintain her weight without using a feeding tube. Months later, she was still spending much of the day in the soft living room chair where she had sat a year earlier before anyone realized she was deficient in a vitamin.

But by 2019, Lisa could dress herself and apply her own mascara. She could push the round button on her electric toothbrush, a feat she worked on for weeks with an occupational therapist. She was nearly able to stand in the shower on her own. "For the first time, I was able to get my walker, step down into the garage, open the car door, back myself up, and get into the seat," she celebrated that summer. Each task she conquered moved her closer to being able to do the thing she missed most: pressing her foot on the gas pedal and setting out on the open road.

Chapter Nine

FILTH PARTIES

IN THE EARLY 1900S, PEOPLE IN THE SOUTH-eastern United States began dying from a strange illness. Over subsequent years, the disease became the deadliest affliction of its kind in American history. Newspapers described the "panic" instilled by the mysterious killer. Three million men, women, and children came down with the disease, and nearly one hundred thousand people died from it. After the condition ravaged the population for twenty-five years, doctors discovered that they could treat it for less than ten cents per patient—and the cure had been at their fingertips all along.

The story begins in rural South Carolina in the summer of 1907, when a middle-aged mother of eight developed a pink rash on her forehead, nose, and cheeks. A few weeks later, her digestive system went awry. Her muscles melted away, leaving her skin hanging like drapes around her arms. She lapsed into a confusion so severe that by the time her husband brought her to the South Carolina State Hospital for the Insane that fall, she spoke in fragments of sentences and could not follow conversations with her children.

Despite her doctor's efforts, the woman's symptoms worsened. She hardly spoke. Even the few words she managed to join together often made little sense. Night and day swirled into a single form, so she might be awake at two or three in the morning and fast asleep from morning until dinnertime. Her energy dwindled. Lifting her limbs became a daunting feat, as if whatever disease afflicted her mind had also sucked the oomph from her flesh and left her with the deflated remnants of a once-hardy body.

Soon after the woman was brought to the asylum, two more patients arrived with the same constellation of symptoms. One was a thirty-year-old housekeeper who had become nearly mute, speaking only when spoken to. In place of sentences, she used syllables that held little meaning in combination. Language eluded her. She was stone-faced at some moments and agitated at others, assaulting people around her at random. A crusty rash had appeared on the backs of her hands and the tops of her feet. Her stomach was violently upset. "If this is a hookworm disease," a doctor wrote of her case, "its symptoms are entirely different from those I am familiar with."

The third patient, a man, came to the asylum around the same time, also with the triad of confusion, upset stomach, and rash. The man spouted strings of jumbled speech dotted with profanities. He fixated on religion, proclaiming everything in the name of Jesus and saying he spoke the word of God. When doctors asked him about his medical history and his family, he had little to offer. He could not remember any of it.

Just twenty days after being admitted to the asylum, the man died there. The women died, too.* Few people realized the trio of cases was like the smoke that billows out of a volcano in advance of an eruption.

Most doctors in the United States had never even heard of the disease that killed the patients. It was an affliction called pellagra, a terrible malady that tended to cause confusion, rashes, and diarrhea. The condition had been named a century earlier, after the Italian word for skin, *pelle*. Experts had long known that it affected people in Europe, but textbooks claimed the United States was immune to it. After all, authorities had argued, almost no cases of the disease had ever been reported in the country.

But as months passed, doctors began to understand that textbooks had been wrong. Pellagra was at their doorstep. By 1912, the disease was the most common cause of death at many asylums in the United States. It spread to prisons and orphanages, killing children and adults alike. It even affected penniless farming families living in their own homes. In a single year, South Carolina alone reported more than thirty thousand cases of pellagra. Forty percent of people who developed the disease went on to die from it, amounting to an astronomical, growing death toll. More than ever, the country needed someone who could control the epidemic. They needed Joseph Goldberger.

---

* The first woman discussed here died three months after arriving at the hospital. The date of death of the second woman, the housekeeper, is not documented.

～～～

In 1914, the U.S. surgeon general called on Dr. Joseph Goldberger to solve the mystery of pellagra. Goldberger was a brainy public health officer who had immigrated from the Carpathian Mountains to the Lower East Side at age nine. His father had opened a small bodega in the city and called on Goldberger to make deliveries to customers around the neighborhood. Lore has it that the young Goldberger would hide books in his jacket as he carried the grocery bags around town. Huddled on tenement stoops or in dim apartment hallways, he would lose himself in the bound pages, sinking into faraway worlds as the sounds of the city continued unheard in the background.

As a public health officer, Goldberger had traveled to Washington, DC, to study a typhoid outbreak, to Mexico City to research malaria, to Mississippi to curb a spate of yellow fever cases, and to New Orleans to combat dengue. Along the way, he had inadvertently contracted yellow fever, typhus, and typhoid himself. Wrapped in blankets or huddled over the toilet, he had been unsurprised by the infections; he knew well that every assignment posed the possibility of bodily discomfort, or worse.

At age thirty-nine, Goldberger was researching diphtheria in Detroit when the U.S. surgeon general asked him to turn his attention to pellagra. "It is undoubtedly one of the knottiest and most urgent problems facing the Service," the official wrote to him.

At the time, a hostile debate was growing around the cause of pellagra. One idea, championed by a stern doctor who had made his name arguing that criminality could be discerned from facial structure, was that pellagra resulted from ingesting a toxin from spoiled corn. The other theory, created by a pompous infectious-disease expert who had discovered the cause of yellow fever, was that pellagra was caused by an infection transmitted by flying bugs.

Most politicians detested the spoiled corn theory, for obvious reasons. To indict corn—and to take its kernels out of the American diet—was to overthrow the entire agricultural economy of the United States. Corn grossed $1.5 billion per year at the time. It was enough, as one official put it, "to cancel the interest-bearing debt of the United States and to pay for the Panama Canal and fifty battleships." The infectious theory of pellagra, on the other hand, allowed the government to blame poor people for their own disease. Pellagra, which primarily affected the disenfranchised, could be caused by the despicable sanitary practices of its own victims.

Poring over transcripts from medical publications and conferences, Goldberger came to believe that the common denominator in pellagra cases was neither spoiled corn nor infection, but rather a nutrient-poor diet. He suspected pellagra was caused by something that was missing from the body, and not by a toxin or infection that had invaded it.

Goldberger pointed to the fact that people who cared for pellagra patients in hospitals and orphanages almost never came down with the affliction themselves. They often spent twenty-four hours a day with pellagra victims, eating the same spoiled corn and being bitten by the same buzzing flies, but caregivers of those with pellagra remained mysteriously resistant to the disease.

Goldberger boarded a train to Virginia in March of 1914, bidding farewell to his wife and children. As weeks wore on, he crisscrossed the southern United States, visiting asylums, prisons, orphanages, and hospitals—all places where pellagra had taken root. In each facility, he took notes on who had developed the disease. He collected menus of what had been served to staff and residents. He even looked for insects that might support the infectious theory of pellagra. He found none.

Instead, Goldberger discovered that people who developed pellagra kept an exceptionally monotonous diet. They tended to eat only a few food items that appeared and reappeared at every meal. "The difference relates to the dietary," Goldberger wrote in a highly publicized paper just a few months after setting out on his trip.

In his first official recommendation for curing pellagra, Goldberger called on the government to increase handouts of milk, eggs, and meat to poor families in at-risk regions. But he had no illusions that anyone would take the advice. Pellagra was a disease of the impoverished, hardly a priority for a government on the brink of world war. Without more concrete, experimental proof—and a solution that would not drain the country's coffers—Goldberger's recommendation failed to precipitate any material change for communities in the center of the pellagra epidemic.

So Goldberger did what he had been trained to do: he set up an experiment to determine the cause of the disease outbreak. He identified two orphanages in Jackson, Mississippi, where yearly outbreaks of pellagra had erupted for more than a decade. Most of the children who had survived carried memories of friends who had not.

Goldberger changed the menu at both orphanages to include more animal products and fewer carbohydrates. Every child received fourteen ounces of milk per day. Eggs accompanied every breakfast. Beans and peas appeared at every lunch. Corn muffins graced the dining hall just once or twice weekly—far less than prior to the intervention.

The results of the study were so striking that Goldberger reported them before the planned two-year timeline ended. In one orphanage, the number of children with pellagra went from one hundred five to one. In the other, cases dropped from sixty-seven to zero. "There has been no pellagra at this institution this year," Goldberger reported proudly of the second facility. Further condemning the infectious theory, he noted that poor sanitary conditions and striking overcrowding—both of which increased the risk of infections—had remained unchanged throughout the experiments. "The conclusion is drawn that pellagra may be prevented by an appropriate diet," he wrote, hoping he had proved his case.

But the scientific community was unmoved. Some researchers criticized Goldberger for not having a control group that continued on the high-corn, low-protein diet. Others pointed out that children were not the only ones to get pellagra, and the study results might not apply to adults. If diet was really at the root of pellagra, others countered, food choices must also be able to cause the disease in healthy people—something Goldberger had yet to demonstrate.

Anticipating the criticism, Goldberger had already begun

his most controversial experiment yet. Pellagra was common in hospitals and orphanages in Mississippi, but the state's prisons—where the diet was surprisingly diverse—had almost no cases of the disease. With support from the governor, Goldberger recruited prisoners to participate in a study to determine whether dietary changes could induce pellagra.

The convicts initially viewed the project as an opportunity. Eighty men volunteered for just twelve spots. The planned diet included biscuits, rice, syrup, sweet potatoes, and grits—a menu that dazzled the prisoners. The governor agreed to pardon all participants, as long as they remained in the study for its full duration. For seven of the men, who were serving life sentences for murder, taking part in the experiment was the only opportunity they would have to return to the free world.

But being in the so-called pellagra squad turned out to be horrifically isolating, and even life threatening. Communication with family and nonparticipating prisoners was cut off to prevent food from being smuggled in and reports of distress from leaking out. One participant was so traumatized that he tried to escape. Another wrote a letter to the governor asking for an early release, calling the experience the worst torture he had ever suffered. When the experiment finally ended and the pellagra squad filed into the Mississippi governor's office to receive their pardons, a reporter found the men "pale, weak, and emaciated, two or three of the number scarcely able to walk." An account of the pardon ran in the *Jackson Daily News* under the title "They Ate Their Way to Freedom."

Of the twelve participants, seven developed suspected pellagra over the course of the study. No cases of the disease appeared among people who had continued eating the typical prison food. Just a few years after beginning his research, Goldberger had proved that dietary changes alone could cure, prevent, and induce pellagra.

For a short time after he published the Mississippi prison data, Goldberger received congratulatory letters from around the world. A scientist on the Nobel Prize nominating committee told him that his work deserved a place in Stockholm. Journalists elevated him to heroic standing and lauded him for remaining "exceedingly modest." Some reporters even speculated that he might go on to cure cancer.

Then a deluge of negative feedback began to flow in. An editorial in the *Baltimore Sun* criticized Goldberger for facilitating the release of prisoners, arguing that experimenting on "noble" people would have avoided "the necessity of setting free those who may be suffering from moral diseases worse than pellagra." A professor at Columbia University accused Goldberger of fabricating his results. A doctor announced at a major medical conference that he would not insult the audience by pretending pellagra was caused by a poor diet. Other doctors, ignoring the practicalities of poverty, argued that pellagra would not have persisted for hundreds of years if it were curable with items that could be purchased at local shops. If the dietary deficiency hypothesis were true, they argued, pellagra victims would have cured themselves.

So Goldberger decided there was only one option left—and an extreme one at that. To persuade his colleagues that pellagra was not infectious, he would have to attempt to catch the disease himself. Along with his wife and fourteen other volunteers—most of them health professionals—he conducted the gold standard of infection studies: he tried to catch pellagra through nearly every body-fluid exposure possible. He hoped the experiments, which later became known as "filth parties," would deliver the death blow to the idea that pellagra was contagious.

On April 25, 1916, Goldberger swabbed the nose and throat of a patient with pellagra, then rubbed the mucus on his own throat and inside his nose. He punctured a vein in the patient's arm, drew up a blood sample, and injected six milliliters of the thick fluid into his own shoulder. He performed the same set of experiments on a coworker who had volunteered for the experiment based on his confidence in Goldberger's research. "*Effects*—" he recorded in a notebook: "Both men felt some soreness and stiffness for a day or two in the muscle into which the blood was injected; otherwise nothing was observed."

Three days later, Goldberger was emboldened. He scraped scales off of the rash of two patients with pellagra, then mixed the flakes with four milliliters of the patients' urine and four milliliters of another pellagra patient's stool. He added wheat flour to solidify the ingredients, then kneaded the mush into the shape of pills. To prevent his stomach juices from inactivating a possible contagion, he premedicated himself with a chemical that made his gastric fluid less acidic. Then he ingested the homemade pills.

For the first few days after the experiment, Goldberger felt gassy and full, but otherwise had few complaints. By the third day, flatulence had devolved into bouts of diarrhea, dragging him to the bathroom several times a day.

Before completely recovering, he repeated his experiments with blood, scales, urine, and stool, using samples from several pellagra patients. Four other volunteers, increasingly convinced of his theories now that he was still alive, joined the experiment as well. Goldberger's wife, a fifth volunteer, joined the protocol at her own insistence. Goldberger refused to let her ingest the stool-laden capsules, but agreed to inject her abdomen with a sample of blood taken from a woman dying of pellagra. For a couple who had spent much of their marriage writing letters to each other from hundreds of miles apart, the joint participation in the filth experiments carried a sort of strange, scientific intimacy.

Other than a tender lymph node and transient diarrhea, Goldberger found no significant negative effects among the volunteers who participated in his seven "filth parties" in the spring of 1916. In the six months after the experiments were completed, not a single volunteer developed pellagra. Several participants suffered temporary upset stomachs, but Goldberger felt this was to be expected. "When one considers the relatively enormous quantities of filth taken," he wrote, "the reactions experienced were surprisingly slight."

Of all the volunteers, Goldberger was the only one to subject himself to all seven experiments. He emerged without any permanent damage. Despite stomach-churning exposures, in a protocol that would never be approved by today's ethicists, he could not catch pellagra.

The critics, however, were not silenced. Some pointed out that Goldberger had only proved that pellagra could not be transmitted from person to person. If the infectious theory was right, pellagra transmission might require an insect intermediary, a possibility that the filth experiments essentially ignored. Other detractors questioned why Goldberger had chosen almost exclusively white men—thought to be the least vulnerable to pellagra—as subjects.

In response, Goldberger decided to go big. Until then, each of his studies on pellagra had dealt with niche groups of people: prisoners, orphans, the mentally insane, and colleagues who had volunteered for his filth parties. To put his critics to rest, he realized, he needed to study the general population.

So for his final experiments in humans, he visited textile-producing villages in rural South Carolina. The enclaves were unique because almost everyone who lived there worked at the mill and was paid with credits that could be used only at the company store. This meant that food choices were limited by what was in stock, and—more importantly—Goldberger could figure out what people were eating by combing the cashiers' ledgers.

Comparing purchases from households that did and did not develop pellagra, Goldberger found a quantitative difference: those without pellagra bought twice as much fresh meat and milk, and four times as much cheese, as households with pellagra. While everyone took home about the same total number of calories, there was something about the type of calories people consumed that determined whether or not a person would develop pellagra. The issue, he argued yet again, was that people with the disease were missing a nutrient from their diet.

With the results of the textile-village experiments, the infectious theory of pellagra finally lost its footing. The man who had championed the idea turned his attention to oncology, arguing unsuccessfully that cancer was caused by cockroaches. In the meantime, Goldberger's data appeared ever more definitive, and at last, most public health experts joined him in believing that pellagra was caused by a dietary deficiency.

~~~~~~

EVEN AS GOLDBERGER BECAME the nationally accepted authority on pellagra, he was still not sure what nutrient was missing from the diet of people with the disease. To find the answer, he turned from human experiments to animal ones, focusing on dogs that had developed a pellagra-like syndrome called blacktongue. Of the fifteen years he would work on the problem of pellagra before his death, nearly half would be spent meticulously modifying dog diets.

The process was tedious and time-consuming, but it eventually paid off. In 1928, Goldberger published a report of sixteen foods he had given to dogs with blacktongue. He showed that maize, carrots, and tomatoes were useless in preventing the disease, but tiny amounts of beef, liver, salmon, and egg yolk could stop blacktongue from developing in even the most malnourished dogs. Searching for a nutrient that might be common and exclusive to the foods that cured blacktongue, Goldberger soon came to suspect that the missing molecule was an as-yet-undiscovered part of the vitamin B complex.

He was right, but he would not live long enough to know it. Goldberger died of kidney cancer in 1929, almost a decade before scientists discovered the identity of the missing vitamin.

He was nominated for a Nobel Prize four times; he never won the award. Posthumously, he became the best-known character in the pellagra story, the subject of books and tales of heroic scientific persistence. His original notes were taken to the National Library of Medicine for safekeeping, and the name "vitamin G" emerged as a temporary eponym for the pellagra-preventing nutrient.

～～～～

Not long after Goldberger died, a biochemist named Conrad Elvehjem picked up the baton. As a quiet farm boy growing up in Wisconsin, Elvehjem had marveled at the way a corn seedling could sprout a cob and kernels seemingly out of thin air, transforming from a nascent plant to a life-sustaining crop in just a few months. His love of agriculture had led to a defining career in nutrition science, so that by the time he tackled the problem of pellagra in the mid-1930s, he was already a recognized name in the field.

In studying Goldberger's work, Elvehjem gleaned that the experiments in both dogs and humans had a fundamental limitation: the deficient diets Goldberger had designed all lacked more than one vitamin. Likewise, the curative foods he had identified also contained multiple vitamins. Without molecular studies, it would be impossible to determine which vitamin in particular was the so-called vitamin G. This was why Goldberger had faltered at the end of his career; he had been dependent on his training in public health, but by that point the problem required a biochemist.

The solution was to take a food that was known to cure

blacktongue—Elvehjem chose liver—and to separate it into its different types of molecules. Then, Elvehjem could check which of the molecules cured blacktongue.

A simplified version of the same experiment can be performed at home. To determine which ingredient makes lemonade sweet, you could heat a cup of lemonade until the water and lemon juice evaporate, then taste the substance left behind. You would find it is sweet. If you went on to identify its chemical structure, you would see you had just isolated sugar. In this way, you could show that sugar—and not water or lemon juice—makes lemonade sweet.

Elvehjem and his colleagues began with four hundred grams of animal liver, then treated the sample with chemicals designed to separate different types of molecules. After a multistep procedure that took several days and left a sinkful of dirty dishes, the scientists produced two grams of a colorless solid that they hoped would treat blacktongue.

The substance behaved like a miracle cure. After ingesting just a small amount of it, dogs with blacktongue enjoyed a near-immediate recovery. Appetites increased. Weight reaccumulated. The dogs became playful again. Within three days of eating the concoction, without any change in diet other than receiving the supplement from Elvehjem's laboratory, it was nearly impossible to tell that the dogs had been near death just seventy-two hours earlier.

Elvehjem went on to find that the curative substance contained carbon, hydrogen, nitrogen, and oxygen, in a ratio that was surprisingly familiar. Atomically speaking, vitamin G looked almost identical to a small molecule called nicotinic acid.

Nicotinic acid had in fact been discovered decades earlier by Casimir Funk, the biochemist who coined the name *vitamine*. Funk had been looking for a cure to multiple neuritis at the time and had been disappointed to find that nicotinic acid did not have any effect on the disease—an unsurprising finding now that we know the condition develops because of a deficiency of thiamine. Although Funk suspected that pellagra was caused by the lack of a vitamin, he did not realize that the cure for the affliction was on hand just across his laboratory. Instead, samples of nicotinic acid sat on his shelf until he retired, and the world waited two decades for Conrad Elvehjem to put the pieces together.

~~~~~~~~

A molecule of nicotinic acid weighs only a fifth as much as a single DNA nucleotide, but it is just as important for cellular survival. We absorb the vitamin in our stomachs and small intestines, then transport it through the bloodstream to the liver—the same organ that Goldberger found to be rich in the pellagra-preventive factor.* In the liver, we transform nicotinic acid into nicotinamide adenine dinucleotide (NAD), a molecule that is essential to the work of at least four hundred proteins—more than any other similar substance in the human body.

NAD is most famous for its role in helping the body reap

---

* We primarily derive nicotinic acid from our diets, but our bodies can make a small amount of the molecule from an amino acid called tryptophan.

energy from sugar molecules. Without it, we struggle to turn food into fuel. Since cells in the skin and gut replicate and divide frequently, and those in the brain are highly metabolically active, all three organ systems are particularly affected by low levels of NAD. This is why pellagra's classic triad of rash, diarrhea, and dementia develops.

But the importance of NAD is hardly limited to its role in digesting sugar. The small molecule is also necessary for repairing damaged DNA. It helps to determine whether stretches of genetic material will be coiled up in tight packages or laid out straight—a pathway thought to be related to aging. Recently, some scientists have even considered whether dietary supplements of NAD might mitigate the effects of time on the body.

Soon after Elvehjem discovered that nicotinic acid cured blacktongue in dogs, doctors began giving the vitamin to patients with pellagra. In North Carolina, a forty-two-year-old farmer who weighed just ninety pounds and had suffered bouts of pellagra for fifteen years received sixty milligrams of nicotinic acid daily. Twenty-four hours after the first dose, the man's appetite returned. He became reoriented, once again able to answer correctly when doctors asked where he was and what day it was. After six days of supplements, the man became "entirely rational." Cracked, scaly skin became springy again by the twelfth day. "The results of this treatment were dramatic," wrote the doctors who cared for the man. "Nicotinic acid is very cheap," they went on; "it cost less than ten cents to treat this patient." Three decades after pellagra first appeared at the South Carolina State Hospital for the Insane, two decades after Joseph

Goldberger injected himself with the blood of a patient with pellagra, Conrad Elvehjem discovered how to treat the disease for the equivalent of $1.80 in today's economy.

Elvehjem's work finally explained the rise and fall of pellagra in the southern United States. Pellagra first became a major problem in the early 1900s as cotton took up more and more agricultural acreage in South Carolina and other nearby states. Without land to grow vegetables, sharecroppers had survived on tithings supplied by landowners—usually corn that had been processed in a way that removed its germ, the main source of nicotinic acid. When a parasite called the boll weevil infiltrated swaths of cotton crops across the region in the late 1910s, sharecroppers began growing their own food on land that was no longer hospitable to cotton. With the change in agricultural practices, pellagra mortality decreased. When the boll weevil receded years later and the cotton economy recovered, the incidence of pellagra soared. Tens of thousands of people dropped dead from the disease, and thousands more suffered disfiguring rashes and horrible bouts of diarrhea and confusion. Finally, during the Great Depression, a time when you might expect pellagra to reach catastrophic numbers, mortality from the disease was instead cut in half. With the value of cotton at an all-time low, sharecroppers had begun growing squash, peas, okra, and other vegetables that easily supplied enough nicotinic acid to prevent the disease from developing.

The final blow to pellagra—and to most vitamin-deficiency diseases in the United States—came in the late 1930s. As World War II became imminent and the thought of bread lines weighed on the American psyche, baking companies across the country collaborated with the U.S. Food and Nutrition Board to set requirements for fortifying flour with nicotinic acid, thiamine, and other vitamins and minerals. The cost of adding the nutrients was negligible; a few cents could get a person enough essential vitamins to last for months. The *New York Times* ran an article titled "Superflour," expounding that "the war will do more for the general consumption of needed vitamins than all the preaching of the nutrition experts."

Bakers working on the nutrition legislation had one major request: they worried the name *nicotinic acid* would confuse buyers, since it sounded like nicotine. The concern was reasonable. Headlines like "Tobacco in Your Bread" had emerged in lay publications. To remedy the confusion, baking companies formed a subcommittee to brainstorm ideas for alternative names for nicotinic acid. They produced *niacin*, *niamin*, and *niacid* as options. *Niacin* won.

By the end of World War II, the incidence of pellagra had plummeted. People across the country consumed niacin regularly in their bread, cereal, cornmeal, and grits. Advances in farming technology decimated the sharecropping industry, so former farmers flocked to the inner city, where they lived off fortified foods. Niacin, the once-powerful molecular evader that threatened the minds of millions, was finally subdued.

Epilogue

The life expectancy of an Alzheimer's disease patient has changed little since the condition was discovered more than a century ago. Aside from improved diagnostic tools and medications that slightly decrease the severity of symptoms, the unfolding of horrors for those with Alzheimer's disease is roughly the same as it was when Alois Alzheimer met Auguste Deter in 1901. Likewise, Huntington's disease still electrifies limbs and muddles minds with the same voracity as before. Creutzfeldt-Jakob disease continues to quickly kill everyone it afflicts, without exception.

But the way we explore cognitive afflictions has begun to change. We know more about the mutants, rebels, invaders, and evaders of the mind than ever before. A century ago, every condition discussed in this book was untreatable. Today, most of the afflictions are preventable. Several are even curable. The field of cognitive neurology has taken its first steps along the path toward molecular specificity. Cognitive calamities are now understood as molecular problems that require molecular solutions.

We have already seen this shift yield results. In 2011, scientists discovered the most common genetic cause of frontotemporal dementia, the condition that caused Danny Goodman to shed empathy and inhibitions and to nearly destroy his wine company (Chapter 3). The DNA mutant has since been implicated in 10 percent of cases of frontotemporal dementia and has been linked to amyotrophic lateral sclerosis (ALS), too. With the help of in vitro fertilization, thousands of people worldwide now have the ability to prevent the mutant gene from appearing in their offspring. Doctors can excise the genetic error from family lineages while still allowing people to have biological children. For those living with the mutation, a clinical trial has even opened using a DNA-like molecule to prevent symptoms from developing.* Less than a decade after discovering the mutation, we already have a candidate cure—a pace of progress previously unheard of.

Research on NMDA receptor encephalitis—the disease that pulled Lauren Kane into the world of *The Walking Dead* (Chapter 4)—has likewise become an exemplary tale of rapid success. More than one hundred cases of the disease were identified in just a single year after researchers discovered its cause. In the past two decades, scientists have described eleven other conditions that are also caused by antibodies that attack molecules floating on the surface of neurons. Each antibody causes a unique constellation of symptoms that neurologists around the world can

---

* The trial is being conducted in those with ALS, since this disease tends to progress faster than frontotemporal dementia and may therefore yield results more quickly in a research study. However, similar trials for those with frontotemporal dementia are beginning soon.

now identify in their patients. With every new antibody discovered, more people receive a molecular explanation of their ills. Most of these patients now go on to a complete recovery.

The conquest of vitamin deficiencies has been one of the biggest molecular triumphs of all. Pellagra was the tenth-leading cause of death in the southern United States in the early twentieth century. Today, there are barely any cases at all in the country. The same is true for thiamine deficiency; patients known to abuse alcohol now receive prophylactic doses of thiamine when they go to the hospital—a cheap and simple way to prevent the disease that caused Lisa Park (Chapter 8) to unknowingly create memories of events that never happened.

The success stories in the preceding pages exist only because neurologists and scientists decided to battle the molecules that conjure cognitive maladies. In many cases, the researchers faced overwhelming criticism from peers. Some presented pivotal results to completely disinterested colleagues. Others were lampooned. Several died before seeing their work come to fruition.

But today, scientists looking to cure dementia receive increasing recognition and financial support. Federal funding for dementia research in the United States increased by $1.7 billion between 2014 and 2019. As of February 2021, the National Institute on Aging was supporting nearly three hundred clinical trials aimed at improving dementia prevention, treatment, and care. The increased resources have been, in part, a reaction to our growing understanding of the astounding scale of cognitive decline. The number of people living with dementia worldwide is expected to triple by 2050, unless we can figure out how to change course.

So here we are, on the precipice of a molecular shift, standing on the proverbial shoulders of the research woven into this book.

~~~~~~~

IN MY OWN CLINIC, I still begin each appointment with long-used tools for cognitive diagnosis. I elicit stories from patients and their families. I review brain scans, eyeing the undulating ripples on the surface of the organ and the tightly packed structures buried inside. I make a diagnosis, giving a rough estimate of how confident I can be that a patient suffers from one disease or another. We discuss prognosis and then medications—the few that yield a subtle improvement in cognition, and the many we use to help manage the depression, anxiety, and agitation that can dominate everyday life for people with dementia.

Then, we talk about the cutting edge. We discuss the option of participating in research. Neurologists used to enroll patients in dementia studies almost blindly, categorizing participants based on symptoms, cognitive testing, and the structures of their brains. The result was often a heterogeneous mix of patients, all being treated with a single drug designed for a disease that many of the participants did not have. At a molecular level, we had no idea what was happening in the brains of people who enrolled.

Now, all of that has changed. The vast majority of clinical trials for dementia are deeply rooted in molecular data. We recruit participants not only based on their symptoms but also according to the molecules we identify in their brains, blood, and spinal fluid. For Alzheimer's disease drug trials, we often use a new type of brain scan to prove that a participant has accumulated the plaques and tangles necessary for a diagnosis of the disease. In frontotemporal dementia, many trials now require genetic testing to prove that a person carries a mutation that is likely to respond to the drug being studied. The same is true for Huntington's disease; drug trials conducted for the affliction now require proof that a participant carries a pathological number of CAG repeats in the Huntington's disease gene. The changes reflect the developing understanding that dementia is not a single disease but rather a symptom that can be caused by many different molecular abnormalities—each deserving its own tailored treatment. We are practicing personalized medicine more than ever before, pitting molecule against molecule as we look to treat horrifically common ailments of the mind.

In twenty-five years, if all goes well, we will look back on the dark days when dementia still meant an irreversible march toward the erasure of the mind. We will tell the story of how we used molecular science to save hundreds of thousands of brains from wilting into nonexistence—and the people we rescued will be there with us, recalling the tale.

ACKNOWLEDGMENTS

I AM INDEBTED TO THE REAL AMELIA ELLMAN, Russell Goodman and his wife, Lauren Kane and her mother, Mike and Amy Bellows, Joe Holloway's wife, and Lisa and Johnny Park, for telling me their stories. Many of them opened their homes to me, and I will fondly remember renting cars and setting out on the road to visit them. I am also grateful to patients who spoke with me but whose accounts I was unable to include here.

I cannot bestow adequate praise on my agent, Steve Ross. He has been a superb advisor on writing and life. I also thank David Doerrer at Abrams Artist Agency, as well as Charlotte Reed. I am grateful to Melanie Tortoroli at W. W. Norton, who tenderly cared both for my sentences and my sanity. I also thank Quynh Do, who always had a vision for this book and pushed me back on track when I needed it. Thanks also to Sarah Johnson for copyediting and to Mo Crist for pulling off all the logistics.

I thank my writing teachers: Linda Press Wulf, for suffering through many drafts of this book and delivering feedback with honesty and compassion; and the Kelly Writers House writing group, for their careful eyes and creative fixes. Thanks also to Sam Apple and Allison LaFave for their suggestions, which were always right.

I thank Dan Kahne and Rahul Kohli for having me in their laboratories. Both are brilliant researchers and wonderful humans.

I am lucky enough to have colleagues who helped me create and fine-tune the medical arc of this book. Many took time out of their clinics and research to help make this text clearer and more precise. For this I thank Geoff Aguirre, Joe Berger, Anjan Chatterjee, Murray Grossman, Dina Jacobs, Francis Jensen, Jason Karlawish, Eric Lancaster, Sanjeev Vaishnavi, and David Wolk. Each of them has taught me to be a better neurologist.

I benefited significantly from advice and discussion with many of the scientists whose work is detailed in this book: Jim Gusella, Norbert Hirschorn, Ken Kosik, Francisco Lopera, Stan Prusiner, and Sonia Vallabh. Thanks also to Alice Wexler for helping me get her family's story right.

I thank my parents, Sue Rodgin and Warren Manning, for their DNA and their edits. Thanks to my in-laws Joan and Martin Peskin, for reading three drafts of this book at record speed without a word of complaint. To Anya Manning, Elie Lehmann, and Isaac Rodgin for their expert reading and their many years of friendship; I lucked out having them for siblings and editors. Thanks to Don Press for stripping out the scientific jargon that I had no clue was in here. I am grateful to Charles and Rita Manning, who I know will put this book on their coffee table even if it is a complete failure everywhere else in the world.

To Jeremy, who has read more versions of this work than anyone else: thank you for your optimism and your confidence in me. To JJ and Ollie, who didn't exist when all this started: you are the biggest joy of our lives. Holding you is bliss. And finally, to Ufruf: you are the best listener I know.

GLOSSARY

Amino acid: The building block of proteins, amino acids contain nitrogen, oxygen, and hydrogen atoms in a standardized arrangement. Branching off of this structure are different "side chains" that determine the specific chemical properties of each amino acid.

Antibody: A protein made by the immune system to help identify and destroy foreign invaders. Antibodies have a stem that binds to molecules from the immune system, and branches that stick like lock and key to a target. Because of this structure, antibodies serve as the physical link between an invading molecule and the immune system.

Antioxidant: A substance that protects cells from dangerous forms of oxygen.

Cell: The smallest units of life, cells are composed of a sloshy substance, called cytoplasm, that is surrounded by a membrane. Neurons are a critical type of cell in the nervous system. The brain is made of billions of cells, each built from molecules, which are constructed from atoms.

Chromosome: A long string of DNA wrapped around supportive proteins. Human DNA is divided into forty-six chromosomes.

Deoxyribonucleic acid (DNA): A molecule composed of nucleotides sewn together in a chain, DNA contains the genetic information that dictates hereditary traits.

Immune system: A complex network of organs, cells, and molecules that has evolved over millions of years to protect the body from foreign invaders.

Molecule: A group of two or more atoms chemically bound together.

Neuron: A specialized cell that receives, processes, and transmits signals, often over long distances. Neurons are the fundamental cell of the nervous system.

Neurotransmitter: A molecule emitted from the end of a nerve and sensed by a neighboring cell, allowing propagation of a signal from one cell to the next.

Nucleotide: The building block of DNA, nucleotides contain phosphorous, oxygen, carbon, nitrogen, and hydrogen atoms. DNA is constructed of four different nucleotides, abbreviated A, T, G, and C, strung together in long chains.

Protein: Molecules built of amino acids strung together and folded into complex, three-dimensional shapes. Proteins are the workhorses of the human cell, responsible for carrying out the basic activities that keep us alive.

Small molecule: A term of art used to refer to molecules that are small enough to move in and out of cells without the help of other molecules.

Vitamin: A molecule essential for normal cellular metabolism that cannot be produced by the body and instead has to be derived from the diet.

NOTES

I DECIDED TO WRITE THIS BOOK IN 2016, WHILE I was a neurology resident at the University of Pennsylvania. As colleagues peeled off into subspecialties, I found myself at an impasse. I was not particularly drawn to one area of neurology over another, but at the same time I worried about being able to maintain a broad-enough knowledge base to practice general neurology.

On the advice of a mentor, I made a list of the diseases that I most wanted to encounter in my clinic. "Try to picture who you want to be sitting on the examination table when you open the door to the clinic room," the teacher advised. As my list grew, I realized that I was drawn to patients who suffered ailments that changed the mind. Every condition I wrote down had a tendency to alter a victim's personality, requiring that doctors navigate not only the technical details of the illness but also the social implications of identity loss.

Looking at my list, the connection to molecular science was also clear. All the conditions I had included were either treatable using precision medicine, or were under research using molecular tools. Neurologists who see patients with these conditions must address both the macro—the person and his social environment—and the micro—the molecule that causes the illness in the first place. Few doctors appreciate this process of distillation, from the holistic treatment of the patient to the painstaking assessment of the molecule, but this has increasingly become part and parcel of cognitive neurology.

In writing this book, I hoped to offer a glimpse into this process of connecting a patient's narrative to the molecules causing the problem. To do this, I interviewed patients, family members, and physicians between 2016 and 2021. I was also lucky enough to receive feedback, either by phone or email, from nearly every living scientist discussed here. I have changed the names of patients, family members, and treating physicians in order to protect patient privacy. In most other respects, I have tried to maintain the details of these stories without alteration, in order to deliver the truth of the experience of being under assault from a terrorizing molecule.

INTRODUCTION

2   *Your brain has more than:* Suzana Herculano-Houzel, "The Remarkable, Yet Not Extraordinary, Human Brain as a Scaled-up Primate Brain and Its Associated Cost," *Proceedings of the National Academy of Sciences of the United States of America* 109, Supplement 1 (2012): 10661–68.

3   *Even DNA, the largest molecule:* Francesco Gentile, Manola Moretti, Tania Limongi, Andrea Falqui, Giovanni Bertoni, Alice Scarpellini, Stefania Santoriello, Luca Maragliano, Remo Proietti Zaccaria, and Enzo di Fabrizio, "Direct Imaging of DNA Fibers: The Visage of Double Helix," *Nano Letters* 12 (2012): 6453–58.

5   *At one end of the room:* Tania J. Lupoli, Tohru Taniguchi, Tsung-Shing Wang, Deborah L. Perlstein, Suzanne Walker, and Daniel E. Kahne, "Studying a Cell Division Amidase Using Defined Peptido-

glycan Substrates," *Journal of the American Chemical Society* 131, no. 51 (2009): 18230–31.

5   *In another corner:* Christine L. Hagan, Seokhee Kim, and Daniel Kahne, "Reconstitution of Outer Membrane Protein Assembly from Purified Components," *Science* 328, no. 5980 (2010): 890–92.

5   *A few desks over:* Shu-Sin Chng, Mingyu Xue, Ronald A. Garner, Hiroshi Kadokura, Dana Boyd, Jonathan Beckwith, and Daniel Kahne, "Disulfide Rearrangement Triggered by Translocon Assembly Controls Lipopolysaccharide Export," *Science* 337, no. 6102 (2012): 1665–68.

Part One: DNA MUTANTS

11   *Enthralled by the chemistry of pus:* Sophie Juliane Veigl, Oren Harman, and Ehud Lamm, "Friedrich Miescher's Discovery in the Historiography of Genetics: From Contamination to Confusion, from Nuclein to DNA," *Journal of the History of Biology* 53, no. 3 (2020): 451–84.

12   *The paper was dry and verbose:* Friedrich Miescher, "Ueber die chemische Zusammensetzung der Eiterzellen," in *Medicinisch-chemische Untersuchungen*, ed. Felix Hoppe-Seyler (Berlin: Verlag von August Hirschwald, 1871), 441–60.

13   *Avery was a nearly retired:* Rene J. Dubos, *The Professor, the Institute, and DNA* (New York: Rockefeller University Press, 1976).

13   *Turning to the research bench:* Nicholas Russell, "Oswald Avery and the Origin of Molecular Biology," *British Journal for the History of Science* 21, no. 4 (1988): 393–400.

Chapter One: IN SUSPENSION

17   *Amelia Ellman sat still:* All information concerning "Amelia Ellman" and her family comes from my interview(s) with the subject.

26   *Then, in 1968:* Alice Wexler, *Mapping Fate: A Memoir of Family, Risk, and Genetic Research* (Berkeley: University of California Press, 1997).

26   *Wexler hosted a series of workshops:* Author interview with Alice Wexler, March 16, 2021.

28   *"It would take more than": An Interview with Nancy Wexler: Chapter 2,* video uploaded April 6, 2020, by Albert and Mary Lasker Foundation, https://www.youtube.com/watch?v=oGXwYDrDJWY; of note, Nancy Wexler recalls scientists predicting that the project would take a hundred years, but transcripts of the meetings reviewed by her sister,

Alice Wexler, suggest that the expected timeline was closer to a decade or two.

29   *In 1993, Wexler and an international team:* J. F. Gusella, N. S. Wexler, P. M. Conneally, S. L. Naylor, M. A. Anderson, R. E. Tanzi, P. C. Watkins, et al., "A Polymorphic DNA Marker Genetically Linked to Huntington's Disease," *Nature* 306, no. 5940 (1983): 234–38.

29   *Huntington's disease turned out to be:* Marcy E. MacDonald, Christine M. Ambrose, Mabel P. Duyao, Richard H. Myers, Carol Lin, Lakshmi Srinidhi, Glenn Barnes, et al., "A Novel Gene Containing a Trinucleotide Repeat That Is Expanded and Unstable on Huntington's Disease Chromosomes," *Cell* 72, no. 6 (1993): 971–983.

30   *Wexler never did:* D. Grady, "Haunted by a Gene," *New York Times*, March 10, 2020.

30   *If the drug is:* For an early-phase clinical trial, see Sarah Tabrizi, Blair Leavitt, Holly Kordasiewicz, Christian Czech, Eric Swayze, Daniel A. Norris, Tiffany Baumann, et al., "Effects of IONIS-HTTRx in Patients with Early Huntington's Disease, Results of the First HTT-Lowering Drug Trial (CT.002)," *Neurology* 90, no. 15 Supplement (2018), https://n.neurology.org/content/90/15_Supplement/CT.002.

Chapter Two: La Bobera de la Familia

33   *Lopera walked into his clinic:* Unless otherwise noted, my account of the case of "Hector Montoya" comes from W. Cornejo, F. Lopera, C. Uribe, and M. Salinas, "Descripcion de una Familia con Demencia Presenil Tipo Alzheimer," *Acta Médica Colombiana* 12, no. 2 (1987), http://actamedicacolombiana.com/anexo/articulos/02-1987-03.pdf.

34   *More than 80 percent:* Alzheimer's Association, "2019 Alzheimer's Disease Facts and Figures," *Alzheimer's & Dementia: The Journal of the Alzheimer's Association* 15, no. 3 (2019): 321–87.

36   *"I have a family here":* Lesley Stahl, "The Alzheimer's Laboratory," *60 Minutes*, aired November 27, 2016, on CBS, https://www.cbsnews.com/news/60-minutes-alzheimers-disease-medellin-colombia-lesley-stahl/.

37   *Dr. Alois Alzheimer was a young:* Gabriele Cipriani, Cristina Dolciotti, Lucia Picchi, and Ubaldo Bonuccelli, "Alzheimer and His Disease: A Brief History," *Neurological Sciences: Official Journal of the Italian Neurological Society and of the Italian Society of Clinical Neurophysiology* 32, no. 2 (2011): 275–79.

38   *Dr. Alois Alzheimer had a habit:* K. Maurer, S. Volk, and H. Ger-

baldo, "Auguste D and Alzheimer's Disease," *Lancet* 349, no. 9064 (1997): 1546–49.

40   *His wife had suddenly become sick:* Nadeem Toodayan, "Professor Alois Alzheimer (1864–1915): Lest We Forget," *Journal of Clinical Neuroscience: Official Journal of the Neurosurgical Society of Australasia* 31 (2016): 47–55.

42   *Alzheimer sliced Auguste's brain:* Toshiki Uchihara, "Silver Diagnosis in Neuropathology: Principles, Practice and Revised Interpretation," *Acta Neuropathologica* 113, no. 5 (2007): 483–99.

43   *Alzheimer began by describing:* Rainulf A. Stelzma, H. Norman Schnitzlein, and F. Reed Murtagh, "An English Translation of Alzheimer's Paper, 'Uber eine eigenartige Erkankung Der Hirnrinde,' " *Clinical Anatomy* 8 (1995): 429–31.

44   *In writing a new edition:* The textbook itself was published in 1910.

46   *The family refused:* Pam Belluck, "A Perplexing Case Puts a Doctor on the Trail of 'Madness,' " *New York Times*, June 2, 2010.

46   *On the way to the church:* Kenneth Kosik, "The Fortune Teller," *The Sciences* 39, no. 4 (1999): 13–17.

46   *"Here you go":* Scene was described by Dr. Kenneth Kosik to the author on March 11, 2021.

47   *Under a microscope:* Francisco Javier Lopera, Mauricio Arcos, Lucia Madrigal, Kenneth S. Kosik, William Cornejo, and Jorge Ossa, "Demencia Tipo Alzheimer con Agregación Familiar en Antioquia, Colombia," *Acta Neurológica Colombiana* 10, no. 4 (1994): 173–87.

48   *He contacted researchers:* Alzheimer's Disease Collaborative Group, "The Structure of the Presenilin 1 (S182) Gene and Identification of Six Novel Mutations in Early Onset AD Families," *Nature Genetics* 11, no. 2 (1995): 219–22.

49   *With the Colombian family's mutation:* Maria Szaruga, Bogdan Munteanu, Sam Lismont, Sarah Veugelen, Katrien Horré, Marc Mercken, Takaomi C. Saido, et al., "Alzheimer's-Causing Mutations Shift Aβ Length by Destabilizing γ-Secretase-Aβn Interactions," *Cell* 170, no. 3 (2017): 443–56.e14.

51   *The medication, which would be given:* Genentech, "Studying Alzheimer's Disease in Colombia," accessed February 27, 2021, https://www.gene.com/stories/our-families-are-waiting.

52   *Underneath was the story of:* Pam Belluck, "Why Didn't She Get Alzheimer's? The Answer Could Hold a Key to Fighting the Disease," *New York Times*, November 4, 2019.

52   *Piedrahita had become an outlier:* Joseph F. Arboleda-Velasquez, Francisco Lopera, Michael O'Hare, Santiago Delgado-Tirado, Claudia

Marino, Natalia Chmielewska, Kahira L. Saez-Torres, et al., "Resistance to Autosomal Dominant Alzheimer's Disease in an APOE3 Christchurch Homozygote: A Case Report," *Nature Medicine* 25, no. 11 (2019): 1680–83.

52 *After her death:* Jennie Erin Smith, "In Life, She Defied Alzheimer's. In Death, Her Brain May Show How," *New York Times*, December 11, 2020.

Chapter Three: HAS ANYONE SEEN MY FATHER?

53 *Danny Goodman had always been:* All information concerning "Danny Goodman" and his family comes from my interview(s) with family members. Additional background about the family business comes from various published articles.

56 *Academically, Pick was raised:* A. Kertesz and P. Kalvach, "Arnold Pick and German Neuropsychiatry in Prague," *Archives of Neurology* 53, no. 9 (1996): 935–38.

57 *Pick eventually became:* M. R. Brown, "Arnold Pick," in *The Founders of Neurology*, 2nd ed., ed. W. Haymaker and F. Schiller (Springfield, IL: Thomas, 1970), 358–62.

57 *He refused to write a textbook:* Dora Fuchs, "Arnold Pick," *Experimental Medicine and Surgery* 9, no. 1 (1957): 192–97.

57 *Just as Auguste Deter prompted:* Arnold Pick, "On the Symptomatology of Left-Sided Temporal Lobe Atrophy," *History of Psychiatry* 8, no. 29 (1997): 149–59.

60 *Pick's work eventually gained the attention:* Mario D. Garrett, "Developing a Modern Mythology for Alzheimer's Disease," *Archives in Neurology & Neuroscience* 4, no. 1 (2019): 1–11.

61 *By 1922, the condition Pick described:* Andrew Kertesz, "Frontotemporal Dementia/Pick's Disease," *Archives of Neurology* 61, no. 6 (2004): 969–71.

62 *Fully aware of his impending death:* One of Pick's students, Dr. Otto Sittig, would carry on Pick's work and publish extensively on the nervous system for two decades, until he was killed in Auschwitz in 1944. The other student, Dr. Erwin Hirsch, became a psychiatrist and fled to Israel at the beginning of World War II.

64 *The result finally answered:* The full story of the genetics of frontotemporal dementia is even more complicated. Twenty percent of cases can be traced back to one of a few genes now known to cause the disease, but eighty percent of cases still have no known genetic cause. If you consider a continuum where one end represents diseases that are uniformly caused by a single gene, and the other end marks conditions that are only rarely attributed to a particular gene, Huntington's disease

and Alzheimer's disease would lie on opposite ends of the spectrum. Frontotemporal dementia would fall somewhere in the middle.

67   *But as is common for people:* A. Tibben, R. Timman, E. C. Bannink, and H. J. Duivenvoorden, "Three-Year Follow-Up after Presymptomatic Testing for Huntington's Disease in Tested Individuals and Partners," *Health Psychology: Official Journal of the Division of Health Psychology, American Psychological Association* 16, no. 1 (1997): 20–35.

Part Two: REBELLIOUS PROTEINS

71   *The history of proteins:* Jamie Wisniak, "Antoine François de Fourcroy," *Revista CENIC Ciencias Químicas* 36, no. 1 (2005): 54–62.

71   *Fourcroy was a counselor:* Arthur C. Aufderheide, *The Scientific Study of Mummies* (Cambridge: Cambridge University Press, 2002).

71   *Over several years, he learned to extract:* Frederic L. Holmes, "Elementary Analysis and the Origins of Physiological Chemistry," *Isis* 54, no. 1 (1963): 50–81.

72   *After a whirlwind career:* J. R. Partington, "Fourcroy. Vauquelin. Chaptal," in *A History of Chemistry* (London: Macmillan Education UK, 1962), 535–66.

74   *The idea was not exactly correct:* Elizaveta Guseva, Ronald N. Zuckermann, and Ken A. Dill, "Foldamer Hypothesis for the Growth and Sequence Differentiation of Prebiotic Polymers," *Proceedings of the National Academy of Sciences of the United States of America* 114, no. 36 (2017): E7460–68.

74   *Scientists have likened proteins:* Charles Tanford and Jacqueline A. Reynolds, *Nature's Robots: A History of Proteins* (New York: Oxford University Press, 2004).

74   *The mass of proteins in our cells:* Ron Milo and Rob Phillips, *Cell Biology by the Numbers* (New York: Garland Science, 2015), 128–31.

Chapter Four: A Zombie Apocalypse

76   *Lauren Kane was born:* Unless otherwise noted, all information about "Lauren Kane" and her family comes from my interviews with her, her mother, and healthcare professionals who treated her, as well as from my own experience with her when I was a neurology resident.

79   *In one clip:* "Lauren" and her mother generously shared this recording with me and gave me consent to reproduce it here.

82   *They collected molecules:* V. H. Maddox, E. F. Godefroi, and R. F. Parcell, "The Synthesis of Phencyclidine and Other 1-Arylcyclohexylamines," *Journal of Medicinal Chemistry* 8, no. 2 (1965): 230–35.

83   *Finally, they infused:* G. Chen, C. R. Ensor, D. Russell, and B. Bohner, "The Pharmacology of 1-(1-Phenylcyclohexyl) Piperidine-HCl," *Journal of Pharmacology and Experimental Therapeutics* 127 (1959): 241–50.

83   *Scientists could then transfer:* M. Johnstone and V. Evans, "Sernyl (Cl-395) in Clinical Anaesthesia," *British Journal of Anaesthesia* 31 (1959): 433–39.

83   *"the most unique compound":* B. S. Nicholas Denomme, "The Domino Effect: Ed Domino's Early Studies of Psychoactive Drugs," *Journal of Psychoactive Drugs* 50, no. 4 (2018): 298–305.

84   *The protein—called the NMDA receptor:* J. W. Newcomer, N. B. Farber, and J. W. Olney, "NMDA Receptor Function, Memory, and Brain Aging," *Dialogues in Clinical Neuroscience* 2, no. 3 (2000): 219–32.

84   *When the tunnel:* Sodium, potassium, and calcium all pass through the tunnel when it is open, but the most important effects are mediated by calcium.

85   *PCP works by sticking:* Lucila Kargieman, Noemi Santana, Guadalupe Mengod, Pau Celada, and Francesc Artigas, "Antipsychotic Drugs Reverse the Disruption in Prefrontal Cortex Function Produced by NMDA Receptor Blockade with Phencyclidine," *Proceedings of the National Academy of Sciences of the United States of America* 104, no. 37 (2007): 14843–48.

85   *"The profundity of PCP":* Marc Lewis, *Memoirs of an Addicted Brain: A Neuroscientist Examines His Former Life on Drugs* (New York: Public Affairs, 2013).

86   *The average person:* L. J. Fanning, A. M. Connor, and G. E. Wu, "Development of the Immunoglobulin Repertoire," *Clinical Immunology and Immunopathology* 79, no. 1 (1996): 1–14.

89   *Days later, Lauren became:* Lauren's recovery was unusually rapid. In many cases, despite optimal treatment, patients take weeks or even months to recover full awareness, since the cells that make the toxic antibody still circulate around the body. Since patients who are treated earlier in the course of the disease tend to recover faster, it is possible that Lauren's rapid improvement happened thanks to her mother's dogged persistence.

89   *What's more, her disease:* Hideto Nakajima, Kiichi Unoda, and Makoto Hara, "Severe Relapse of Anti-NMDA Receptor Encephalitis 5 Years after Initial Symptom Onset," *eNeurologicalSci* 16 (2019): 100199.

Chapter Five: THE MUSCLE MAN

91 **Mike Bellows met Amy Holmes:** Unless otherwise noted, all information about "Mike Bellows" comes from my interviews with him and "Amy Holmes."

97 **thin end of a whip:** W. H. Mcmenemey, "Santiago Ramón y Cajal (1852–1934)," *Proceedings of the Royal Society of Medicine* 46, no. 3 (1953): 173–77.

97 **a single, continuous unit:** Gordon M. Shepherd, "The Neuron Doctrine," *Foundations of the Neuron Doctrine*, 2015.

98 **give up hiring a nanny:** R. Yuste, "The Discovery of Dendritic Spines by Cajal," *Frontiers in Neuroanatomy* 9, no. 18 (2015).

99 **a conference in Berlin:** C. G. Goetz, "Minds Behind the Brain: A History of Brain Pioneers and Their Discoveries," *JAMA: The Journal of the American Medical Association* 284, no. 8 (2000).

100 **a German neurologist named Otto Loewi:** An Mccoy and Sy Tan, "Otto Loewi (1873–1961): Dreamer and Nobel Laureate," *Singapore Medical Journal* 55, no. 01 (2014).

100 **came to him in a dream:** Elliot S. Valenstein, "The Discovery of Chemical Neurotransmitters," *Brain and Cognition* 49 (2002): 73–95.

Chapter Six: DEADLY LAUGHTER

106 **He had been eager:** Vincent Zigas, *Laughing Death: The Untold Story of Kuru* (New York: Humana Press, 1990).

110 **"They're all dead":** Quote has been paraphrased; original is found in Zigas, *Laughing Death*.

117 **In 1921, a German neurologist:** A. Jakob, "Über eigenartige erkrankungen des zentralnervensystems mit bemerkenswertem anatomischen befunde (Spastische pseudosklerose—encephalomyclopathie mit disseminirrten degenerationsherden)," *Zeitschrift für die gesamte Neurologie und Psychiatrie* 64 (1921): 147–228.

117 **Creutzfeldt's case turned out:** F. Katscher, "It's Jakob's Disease, Not Creutzfeldt's," *Nature* 393, no. 6680 (1998): 11; and Michael Illert and Mathias Schmidt, "Hans Gerhard Creutzfeldt (1885–1964) in the Third Reich: A Reevaluation," *Neurology* 95, no. 2 (2020): 72–76.

118 **The veterinarian wrote a letter:** W. J. Hadlow, "Scrapie and Kuru," *Lancet* (1959): 289–90.

120 **He moved to Australia:** D. C. Gajdusek, "Vincent Zigas 1920–1983," *Neurology* 33, no. 9 (1983): 1199.

121   *Researchers could heat:* P. Brown, P. P. Liberski, A. Wolff, and D. C. Gajdusek, "Resistance of Scrapie Infectivity to Steam Autoclaving after Formaldehyde Fixation and Limited Survival after Ashing at 360 Degrees C: Practical and Theoretical Implications," *Journal of Infectious Diseases* 161, no. 3 (1990): 467–72.

121   *Dr. Stanley Prusiner was:* S. B. Prusiner, *Madness and Memory: The Discovery of Prions—A New Biological Principle of Disease* (New Haven, CT: Yale University Press, 2016).

122   *He coined the word:* S. B. Prusiner, "Novel Proteinaceous Infectious Particles Cause Scrapie," *Science* 216, no. 4542 (1982): 136–44.

123   *Every human in the world:* R. A. Maddox, M. K. Person, J. E. Blevins, J. Y. Abrams, B. S. Appleby, L. B. Schonberger, and E. D. Belay, "Prion Disease Incidence in the United States: 2003–2015," *Neurology* 94, no. 2 (2019): e153–e157.

123   *In a series:* K. M. Pan, M. Baldwin, J. Nguyen, M. Gasset, A. Serban, D. Groth, I. Mehlhorn, et al., "Conversion of Alpha-Helices into Beta-Sheets Features in the Formation of the Scrapie Prion Proteins," *Proceedings of the National Academy of Sciences of the United States of America* 90, no. 23 (1993): 10962–66.

123   *Like a room full of:* Jay Ingram, *Fatal Flaws: How a Misfolded Protein Baffled Scientists and Changed the Way We Look at the Brain* (New Haven, CT: Yale University Press, 2013).

124   *In 2019, scientists showed:* E. V. Minikel, H. T. Zhao, J. Le, J. O'Moore, R. Pitstick, S. Graffam, and G. A. Carlson, "Prion Protein Lowering Is a Disease-Modifying Therapy across Prion Disease Stages, Strains and Endpoints," *Nucleic Acids Research* 48, no. 19 (2020): 10615–31.

124   *Prusiner has begun to argue:* Joel C. Watts and Stanley B. Prusiner, "B-Amyloid Prions and the Pathobiology of Alzheimer's Disease," *Cold Spring Harbor Perspectives in Medicine* 8, no. 5 (2018): a023507.

124   *Most scientists still consider:* Alison Abbott, " 'Transmissible' Alzheimer's Theory Gains Traction," *Nature*, December 13, 2018, https:// doi.org/10.1038/d41586-018-07735-w.

Part Three: BRAIN INVADERS AND EVADERS

129   *We witnessed the torture:* Charles Tanford and Jacqueline A. Reynolds, *Nature's Robots: A History of Proteins* (New York: Oxford University Press, 2004).

130   *The idea of molecular evaders:* Kenneth J. Carpenter, "A Short His-

tory of Nutritional Science: Part 1 (1785–1885)," *Journal of Nutrition* 133 (2003): 638–45.

132 *Funk himself nearly became:* Paul Griminger, "Casimir Funk," *Journal of Nutrition* 102, no. 9 (1972): 1105–13.

133 *Use of medicinal plants:* K. Hardy, "Paleomedicine and the Evolutionary Context of Medicinal Plant Use," *Revista Brasileira de Farmacognosia* 31 (2020), https://doi.org/10.1007/s43450-020-00107-4.

133 *Documents written in 1550 BCE:* Cyril P. Bryan, trans., *The Papyrus Ebers* (New York: D. Appleton, 1931).

133 *Even the first:* Alan Wayne Jones, "Early Drug Discovery and the Rise of Pharmaceutical Chemistry," *Drug Testing and Analysis* 3, no. 6 (2011): 337–44.

133 *Today, half of all people:* National Center for Health Statistics, *National Health and Nutrition Examination Survey* (2018), table 38.

134 *Even the active ingredient:* Shelly L. Gray, Melissa L. Anderson, Sascha Dublin, Joseph T. Hanlon, Rebecca Hubbard, Rod Walker, Onchee Yu, Paul K. Crane, and Eric B. Larson, "Cumulative Use of Strong Anticholinergics and Incident Dementia: A Prospective Cohort Study," *JAMA Internal Medicine* 175, no. 3 (2015): 401–7.

Chapter Seven: LIKE LUCIFER

135 *Brawling in a local drugstore:* Herbert Mitgang, "The Law; Lincoln as Lawyer: Transcript Tells Murder Story," *New York Times*, February 10, 1989, https://www.nytimes.com/1989/02/10/nyregion/the-law-lincoln-as-lawyer-transcript-tells-murder-story.html.

136 *According to one courtroom witness:* Gary Ecelbarger, *The Great Comeback: How Abraham Lincoln Beat the Odds to Win the 1860 Republican Nomination* (New York: Thomas Dunne Books, 2008).

137 *"as he was grand":* Michael Burlingame, *The Inner World of Abraham Lincoln* (Urbana-Champaign: University of Illinois Press, 1997).

137 *During the fourth debate:* Ward Hill Lamon, *Recollections of Abraham Lincoln, 1847–1865,* ed. Dorothy Lamon Teillard (Lincoln: University of Nebraska Press, 1994).

138 *In moments of uncontrolled wrath:* N. Hirschhorn, R. G. Feldman, and I. A. Greaves, "Abraham Lincoln's Blue Pills: Did Our 16th President Suffer from Mercury Poisoning?," *Perspectives in Biology and Medicine* 44, no. 3 (2001): 315–32.

138 *"His bowels don't ever":* Gore Vidal, *Lincoln: A Novel* (New York: Random House, 1984).

139   *In 241 BCE:* Lydia Kang and Nate Pedersen, *Quackery: A Brief History of the Worst Ways to Cure Everything* (New York: Workman Publishing, 2017).

140   *In the early 1800s:* Marissa Fessenden, "How to Reconstruct Lewis and Clark's Journey: Follow the Mercury-Laden Latrine Pits," *Smithsonian*, September 8, 2015; and Montana Fish, Wildlife and Parks, "Travelers' Rest State Park," accessed February 28, 2021, https://fwp.mt .gov/stateparks/travelers-rest/.

141   *From the respiratory tract:* Robin A. Bernhoft, "Mercury Toxicity and Treatment: A Review of the Literature," *Journal of Environmental and Public Health* (2012): 460508.

141   *The brain is mercury's most vulnerable target:* David R. Wallace, Elizabeth Lienemann, and Amber N. Hood, "Clinical Aspects of Mercury Toxicity," in *Clinical Neurotoxicology*, ed. Michael Dobbs (Philadelphia: Elsevier's Health Science, 2009), 251–58.

141   *It changes the balance of calcium:* Coral Sanfeliu, Jordi Sebastia, Rosa Cristofol, and Eduardo Rodriquez-Farre, "Neurotoxicity of Organomercurial Compounds," *Neurotoxicity Research* 5, no. 4 (2003): 283–306.

141   *Children exposed to mercury:* J. Warkany and D. M. Hubbard, "Adverse Mercurial Reactions in the Form of Acrodynia and Related Conditions," *A.M.A. American Journal of Diseases of Children* 81, no. 3 (1951): 335–73.

141   *In 1988, a botched:* R. E. Bluhm, R. G. Bobbitt, L. W. Welch, A. J. Wood, J. F. Bonfiglio, C. Sarzen, A. J. Heath, and R. A. Branch, "Elemental Mercury Vapour Toxicity, Treatment, and Prognosis after Acute, Intensive Exposure in Chloralkali Plant Workers. Part I: History, Neuropsychological Findings and Chelator Effects," *Human & Experimental Toxicology* 11, no. 3 (1992): 201–10; and *Olin Corporation v. Yeargin Incorporated*, 146 F.3d 398 (6th Cir. 1998), Findlaw. com, accessed February 28, 2021, https://caselaw.findlaw.com/us-6th -circuit/1455092.html.

142   *In 2019, less than:* D. D. Gummin, J. B. Mowry, M. C. Beuhler, D. A. Spyker, D. E. Brooks, K. W. Dibert, L. J. Rivers, N. P. T. Pham, and M. L. Ryan, "2019 Annual Report of the American Association of Poison Control Centers' National Poison Data System (NPDS): 37th Annual Report," *Clinical Toxicology* 58, no. 12 (2020): 1360–1541, https://doi.org/10.1080/15563650.2020.1834219.

142   *Most cases of modern-day:* U.S. Environmental Protection Agency, "How People Are Exposed to Mercury," accessed April 5, 2021, https:// www.epa.gov/mercury/how-people-are-exposed-mercury.

143   *The paste was then rolled:* Kate Haddock, *Mystery Files: Abraham Lincoln*, video, 24 min., 2010.

146   *"The most fun about my various researches":* Dr. Norbert Hirschhorn, email to author, July 15, 2019.

Chapter Eight: An Honest Liar

147   *Lisa Park's memory had acquired a curious flaw:* Unless otherwise noted, all information about "Lisa Park" comes from my interviews with her and her husband.

151   *Dr. Sergei Sergeievich Korsakoff was born:* S. Katzenelbogen, "Sergei Sergeivich Korsakov (1853–1900)," in *The Founders of Neurology; One Hundred and Thirty-Three Biographical Sketches*, ed. Webb Haymaker (Springfield, IL: Charles C Thomas, 1902), 311–14.

151   *A burly and serious-looking man:* Alla Vein, "Sergey Sergeevich Korsakov (1854–1900)," *Journal of Neurology* 256, no. 10 (2009): 1782–83.

152   *Korsakoff first met Arkady Volkov:* This is a fictional name used here for clarity; the true name of this patient is unpublished.

154   *With the benefit of an autopsy:* S. S. Korsakoff, "Medico-Psychological Study of a Memory Disorder," *Consciousness and Cognition* 5, no. 1/2 (1996): 2–21.

154   *The constellation of symptoms:* N. J. Arts, S. J. Walvoort, and R. P. Kessels, "Korsakoff's Syndrome: A Critical Review," *Neuropsychiatric Disease and Treatment* 13 (2017): 2875–90.

155   *Around the time:* Nobel Media, "Christiaan Eijkman—Nobel Lecture: Antineuritic Vitamin and Beriberi," NobelPrize.org, accessed February 28, 2021, https://www.nobelprize.org/prizes/medicine/1929/eijkman/lecture/.

155   *But none of the toxins:* K. Pietrzak, "Christiaan Eijkman (1856–1930)," *Journal of Neurology* 266, no. 11 (2019): 2893–95.

157   *Within days of switching back:* C. Eijkman, *Polyneuritis in Chickens, or the Origins of Vitamin Research*, trans. D. G. van der Heij (Basel: Hoffman-la Roche, 1990).

157   *A few years later, in the same institution:* B. C. P Jansen and W. F. Donath, "On the Isolation of the Anti-Beri-Beri Vitamin," *Proceedings of the Royal Academy Amsterdam* 29 (1926).

158   *Thanks to research that:* Lawrence J. Machlin, introduction to *Beyond Deficiency: New Views on the Function and Health Effects of Vitamins*, Annals of the New York Academy of Sciences 669 (1992).

158   *Thiamine is a small molecule:* Peter R. Martin, Charles K. Singleton, and Susanne Hiller-Sturmhofel, "The Role of Thiamine Deficiency in Alcoholic Brain Disease," *Alcohol Research & Health* 27, no. 2 (2003): 134–42.

158   *It is no surprise:* M. D. Kopelman, "Frontal Dysfunction and Memory Deficits in the Alcoholic Korsakoff Syndrome and Alzheimer-type Dementia." *Brain: A Journal of Neurology* 114 (Pt 1A, 1991), 117–37.

158   *Thiamine also protects us:* Alan S. Hazell, "Astrocytes Are a Major Target in Thiamine Deficiency and Wernicke's Encephalopathy," *Neurochemistry International* 55, no. 1–3 (2009): 129–35.

Chapter Nine: FILTH PARTIES

161   *The story begins in rural: What Are Pellagra and Pellagrous Insanity? Does Such a Disease Exist in South Carolina, and What Are Its Causes? An Inquiry and a Preliminary Report to the South Carolina State Board of Health*, December 30, 1907.

163   *Forty percent of people:* Alan M. Kraut, *Goldberger's War: The Life and Work of a Public Health Crusader* (New York: Hill & Wang, 2003).

165   *The other theory, created:* Louis W. Sambon, "Remarks on the Geographical Distribution and Etiology of Pellagra," *British Medical Journal* 2, no. 2341 (1905): 1272–74.

165   *It was enough: Annual Report of the Trade and Commerce of Chicago for the Year Ended December 31, 1909.*

166   *"The difference relates to the dietary":* Joseph Goldberger, "The Etiology of Pellagra: The Significance of Certain Epidemiological Observations with Respect Thereto," *Public Health Reports* 29, no. 26 (1914): 1683–86.

166   *He identified two orphanages:* Joseph Goldberger, C. H. Waring, and David G. Willets, "The Prevention of Pellagra: A Test of Diet among Institutional Inmates," *Public Health Reports* 30, no. 43 (1915): 3117–31.

169   *An editorial in the* Baltimore Sun: "The Reported Conquest of Pellagra," *Baltimore Sun*, November 3, 1915.

169   *A professor at Columbia University:* Mary Farrar Goldberger, "Dr. Joseph Goldberger: His Wife's Recollections," presented before the 38th annual meeting of the American Dietetic Association in St. Louis on October 20, 1955.

169   *Other doctors, ignoring the practicalities:* Elizabeth W. Etheridge, *The Butterfly Caste: A Social History of Pellagra in the South* (Westport, CT: Praeger, 1972).

169   *Along with his wife:* Joseph Goldberger, "The Transmissibility of Pellagra: Experimental Attempts at Transmission to the Human Subject," *Public Health Reports* 31, no. 46 (1916): 3159–73.

173   *In 1928, Goldberger published:* Joseph Goldberger, G. A. Wheeler, R. D. Lillie, and L. M. Rogers, "A Study of the Blacktongue-Preventive Action of 16 Foodstuffs, with Special Reference to the Identity of Blacktongue of Dogs and Pellagra of Man," *Public Health Reports* 43, no. 23 (1928): 1385–1454.

174   *As a quiet farm boy:* R. H. Burris, C. R. Baumann, and Van R. Potter, "Conrad Arnold Elvehjem: 1901–1962," *National Academy of Sciences* (1990).

175   *Elvehjem and his colleagues:* C. J. Koehn Jr. and Conrad Elvehjem, "Further Studies on the Concentration of the Antipellagra Factor," *Journal of Biological Chemistry* 18, no. 3 (1937): 693–99.

175   *Elvehjem went on to find:* C. A. Elvehjem, Robert J. Madden, F. M. Strong, and D. W. Woolley, "The Isolation and Identification of the Anti-Black Tongue Factor," *Journal of the American Chemical Society* 59 (1937).

176   *In the liver, we transform:* Lee Russell McDowell, "Niacin," in *Vitamins in Animal and Human Nutrition* (Ames: Iowa State University Press, 2008), 347–83.

177   *But the importance of NAD:* Peter Belenky, Katrina L. Bogan, and Charles Brenner, "NAD+ Metabolism in Health and Disease," *Trends in Biochemical Sciences* 32, no. 1 (2006): 12–19.

177   *Recently, some scientists:* Eric Verdin, "NAD+ in Aging, Metabolism, and Neurodegeneration," *Science* 350, no. 6265 (2015): 1208–13.

177   *In North Carolina:* David T. Smith, "Pellagra Successfully Treated with Nicotinic Acid: A Case Report," *Journal of the American Medical Association* 109, no. 25 (1937): 2054.

178   *With the value of cotton:* Etheridge, *The Butterfly Caste.*

179   *The New York Times:* "Superflour," *New York Times,* January 12, 1941, E8.

179   *Headlines like:* "Niacin and Nicotinic Acid," *JAMA: The Journal of the American Medical Association* 118, no. 10 (1942): 823.

Epilogue

180   *In 2011, scientists:* Julie van der Zee, Ilse Gijselinck, Lubina Dillen, Tim Van Langenhove, Jessie Theuns, Sebastiaan Engelborghs, Stephanie Philtjens, et al., "A Pan-European Study of the C9orf72 Repeat

Associated with FTLD: Geographic Prevalence, Genomic Instability, and Intermediate Repeats," *Human Mutation* (2012).

181   *For those living with the mutation:* "Biogen-C9 Phase I Clinical Trial," University of Miami ALS Center, accessed February 28, 2021, https://www.miami-als.org/study/biogen-c9-phase-i-clinical-trial/.

182   *Pellagra was the:* Karen Clay, Ethan Schmick, and Werner Troesken, *The Rise and Fall of Pellagra in the American South* (Cambridge, MA: National Bureau of Economic Research, 2017).

182   *The increased resources have been:* World Health Organization and Alzheimer's Disease International, *Dementia: A Public Health Priority* (Geneva: WHO, 2012).

INDEX

Page numbers beginning with 191 refer to endnotes.